Synthesis Lectures on Engineering, Science, and Technology

The focus of this series is general topics, and applications about, and for, engineers and scientists on a wide array of applications, methods and advances. Most titles cover subjects such as professional development, education, and study skills, as well as basic introductory undergraduate material and other topics appropriate for a broader and less technical audience.

Chris Fields · James Glazebrook

Distributed Information and Computation in Generic Quantum Systems

 Springer

Chris Fields
Allen Discovery Center
Tufts University
Medford, MA, USA

James Glazebrook
Department of Mathematics and Computer
Science
Eastern Illinois University
Charleston, IL, USA

ISSN 2690-0300   ISSN 2690-0327 (electronic)
Synthesis Lectures on Engineering, Science, and Technology
ISBN 978-3-031-97262-1   ISBN 978-3-031-97263-8 (eBook)
https://doi.org/10.1007/978-3-031-97263-8

Preface: What This Book is About

Despite Niels Bohr's reservations and indeed warnings about the idea that atomic theory could be *mechanical*, quantum mechanics has been called "quantum *mechanics*" since its initial formal development in the 1920s. Why this insistence on mechanics? Did Heisenberg, Schrödinger, Born, and even Bohr use "mechanics" to get physicists not familiar with the new theory to take it seriously? Was it, in other words, just a pedagogical tactic? In the 1970s, when the two of us were in graduate school, courses and their textbooks were still titled *Quantum Mechanics* and still presented the theory as a "quantization" of the classical mechanics of billiard balls and water waves. The 1970s were the golden age of "shut up and calculate" and the question of what quantum mechanics meant was generally left to the philosophers. Feynman famously said that nobody understands quantum mechanics. If there is, as Bohr put it in his 1913 paper introducing quantized electron orbitals, *no question of a mechanical foundation* for the theory, how could they?

The idea that quantum theory is not about mechanics or even objects, but instead about *information*, was probably first enunciated by Wheeler. His "it from bit" is catchy as a slogan, but what exactly does *from* mean? Is it the straightforward *from* of logical inference, or the much more mysterious *from* of metaphysics? Wheeler characterized quantum systems as "observer-participants" but refrained from mixing quantum theory with consciousness in the ways that von Neumann or Wigner had. He hinted that an "it" is an inference, but never—to our knowledge—came out and said that what we call "objects" are all *just* inferences. In the absence of clarity, a tendency toward metaphysics easily takes over, generating ever more fanciful "interpretations" of the quantum formalism and ever more seemingly desperate attempts to maintain a metaphysically classical world against the onslaughts of Bell's theorem.

Quantum information came into its own as a theory in the 1990s, and was rapidly adopted as the language of both quantum computing and quantum gravity. The development of quantum computers and algorithms is now an enormously well-funded engineering discipline, while quantum complexity and quantum gravity remain arcane, and increasingly mutually entangled, theoretical endeavors. Outside of these specialties, however, the association between "quantum" and "microscopic" is still pervasive, and the idea that something weird but at least vaguely mechanical is going on down in the "quantum world" remains strong.

That quantum theory describes everyday reality—all of it, all the time—remains, in other words, by far a minority opinion.

Our goal in this book is to change that, or at least to make a small contribution toward the sea change in our thinking that is needed to recognize quantum theory for what it is: a fundamental, principled description of both ourselves and everything we experience. *We ourselves* are generic quantum systems, we are surrounded by generic quantum systems, and our interactions with the people and objects around us are generic quantum interactions. Quantum theory is not arcane. It is a conceptually straightforward description of how we and everything around us works. It does not need an interpretation, and it does not need a metaphysics. Indeed it asks us only to abandon a metaphysics, a metaphysics we constructed by taking some of the cheap heuristics with which evolution equipped us to be Laws of Nature.

We approach our task by interleaving the languages of physics and information theory—particularly the theory of distributed computation—with the formal language of category theory. We introduce the idea of physics as a way of talking about information flow in Chap. 1 and use this idea to develop the basics of quantum theory in Chap. 2. We then put this into category-theoretic language in Chap. 3, and use this to recast physics as Bayesian inference in Chap. 4. Here we introduce the *Free Energy Principle* (FEP) developed over two decades by Karl Friston and colleagues, which anchors the inferential view of physics in evolutionary biology and neuroscience. We discuss the context dependence of information flows in Chap. 5, and use this to develop a generic, scale-free quantum formulation of the FEP in Chap. 6. We explore the consequences of these formal moves for the description of multi-agent communication in Chap. 7 and for the description of learning, adaptation, and self-organization in Chap. 8. We conclude in Chap. 9 that physics, the theory of computation, and even biology and the cognitive sciences are just different languages for describing the same phenomena: exchanges of information between generic quantum systems. Various items of formal background are provided as appendices.

The presentation here draws on work with numerous colleagues. In particular, we thank Karl Friston, Michael Levin, Antonino Marcianò, and Emanuele Zappala, with whom many of the ideas presented here were developed. We also thank Ander Aguirre, Ion Baianu, Ronald Brown, Eric Dietrich, Shanna Dobson, Maurice Dupré, Danko Georgiev, Adam Goldstein, Don Hoffman, John McLennan, Timothy Porter, Lars Sandved-Smith, Rodrick Wallace, and many others for discussions of various topics covered here.

Caunes-Minervois, France   Chris Fields
Charleston, USA   James Glazebrook
May 2025

Contents

Introduction

1

Let us start with the basics: what is a "system" and what does it mean to characterize a system as "generic" and "quantum"? What is "information" and how can it be "distributed" between systems? What, finally, does it mean to say that some physical process is a "computation"? This book addresses these questions, and explores the consequences of answering them while making as few assumptions as possible. It shows how answering these simple questions leads to a deep theory with applications throughout physics, and beyond physics into the life and social sciences. This theory is universal, applying to all physical systems at all scales of organization or complexity. It is also scale-free, describing systems at all scales with the same formalism and in the same way. It achieves this by formalizing all physical interactions as information exchange.

1.1 From the Physics of Objects to the Physics of Information

From its origins in antiquity, physics was the study of objects in motion. The Renaissance achievements of Galileo, Kepler, Descartes, and Leibniz all concerned objects in motion. Newton's laws characterize objects in motion. The study of objects in motion had obvious motivations, and equally obvious practical applications. The industrial revolution in Europe was a direct consequence of understanding how objects moved, and how their motions could be controlled.

The study of objects in motion culminated in the 19th century with two abstractions of the intuitive notions of "object" and "motion". The first was Faraday's idea of a "field" pervading space, and its formalization by Maxwell's equations of electromagnetism. Maxwell's equations recast light as an interaction between abstract electric and magnetic fields, and provoked the ultimately fruitless search for a non-abstract, elastic material—the hypoth-

© The Author(s), under exclusive license to Springer Nature Switzerland AG 2026   1
C. Fields and J. Glazebrook, *Distributed Information and Computation in Generic Quantum Systems*, Synthesis Lectures on Engineering, Science, and Technology,
https://doi.org/10.1007/978-3-031-97263-8_1

esized aether—that vibrated or undulated as light traversed it. Light has, indeed, become more abstract, not less, at every major step in the development of modern physics, from the early 20th century to the present day.

The second abstraction was the notion of heat, a diffuse energy that could cause changes in temperature, volume, and pressure. Heat was obviously related to temperature, but depended also on some unknown property of the object being heated. Clausius resolved the problem in 1867 by giving this unknown property a name, writing $d\mathbf{E} = T\,d\mathbf{S}$, where $\mathbf{E}$ is the energy of heat, T is temperature, and $\mathbf{S}$ is an entirely new concept, entropy [1]. Like the notion of a field, entropy was an abstraction. No known property of any object fit the bill.

The idea that would, a century later, give rise to a new physics of quantum information was first enunciated by Boltzmann in 1877. Boltzmann was a proponent of the then-unpopular theory of atoms—a rediscovery of the ancient idea of Democritus that matter comes in indivisible chunks—as the only sensible explanation of the stream of new results coming from chemistry. He reasoned that gross, macroscopic measurements of temperature, pressure, or volume could not precisely specify the positions or motions of all of the atoms in a container of gas or liquid. He recognized, in other words, that measurements at one scale could be ambiguous about—fail to distinguish between—distinct possible states of affairs at a different scale. In a stroke of what can only be called genius, he proposed that this ambiguity of measurement is what Clausius had called *entropy*, writing (in more modern notation) $d\mathbf{S} = k_B \ln\Omega$, where k_B, Boltzmann's constant, has units of $\mathbf{E}/T$ and Ω is the number of states of the system of interest that are indistinguishable by some measurement of interest [2, 3].

It is difficult to overstate either the importance or the revolutionary character of Boltzmann's equation $d\mathbf{S} = k_B \ln\Omega$. With this equation, Boltzmann recognized that macroscopic observables like temperature, pressure, and volume were all themselves abstractions: coarse-grained, ambiguous, instrumental descriptions of an observationally inaccessible underlying reality. Boltzmann's equation thus introduced, albeit implicitly, the observer and the resolution of the observer's measurement apparatus into physics, though it would not be until the development of the theory of quantum reference frames—discussed in Chap. 2, Sect. 2.4 below—in the 1970s that the observer-dependence of the concept of entropy would become widely recognized. By equating $d\mathbf{S}$, a physical quantity, with an observer's inability to distinguish between Ω distinct states of a physical system, Boltzmann also made "uncertainty", and hence the Laplacian or "subjective" notion of probability [4], a fundamental concept of physics. The statistical mechanics that grew out of Boltzmann's work is, effectively, a calculus of probabilities of measurement outcomes.

Boltzmann's most consequential realization, however, was that the reduction of uncertainty is both costly and quantized. The number of indistinguishable states, Ω, can be reduced by no less than one state at a time. The minimum number of states about which an observer can be uncertain is two: up or down, heads or tails, black or white. Obtaining and recording a datum that answers a single yes/no question resolves the uncertainty. *Recording* the datum in a thermodynamically irreversible way—writing it down—is critical to the process of uncertainty reduction, as the written record constitutes retrievable evidence that the particular answer was obtained. Setting $\Omega = 2$ and combining Boltzmann's equation with that of Clausius yields $d\mathbf{E} = \ln 2 k_B T$, as the energy required to answer a single yes/no question. This is Landauer's Principle, which can also be written $d\mathbf{E} = \beta k_B T$, where $\beta \geq \ln 2$ in recognition of the fact that such questions can be answered inefficiently. First stated explicitly in 1961 [5], Landauer's Principle established the per-bit cost of irreversibly-recorded classical information—answers to yes/no questions—as a fundamental fact about the universe [6–8].

It would require the work of Planck, Einstein, and Heisenberg to finally understand that observers could trade off the energy spent making a measurement against the time allowed to complete it, and hence that what is quantized at the fundamental level is not energy $d\mathbf{E}$, but rather action $d\mathbf{E}dt$. The fact that measurements take time is, of course, obvious in practice, but is abstracted away in classical physics, where in the absence of dissipative forces, interactions are represented as instantaneous. Viewed from the perspective of Landauer's Principle, taking more time is equivalent to lowering the temperature. This equivalence is formalized by the Wick rotation, $\iota t/\hbar \leftrightarrow 1/(k_B T)$, a fundamental link between quantum and classical physics not stated explicitly until 1954 [9] and still widely regarded as a "trick" without evident physical meaning. We will discuss it in the context of a formal theory of measurement in Chap. 2 below.

Boltzmann did not, therefore, have all of the ingredients of quantum theory in 1877, but he was tantalizingly close. His 1906 suicide prevented his seeing Bohr's 1913 model of the Hydrogen atom, or reading Bohr's prescient statement that "there obviously can be no question of a mechanical foundation" for his quantized model of electron orbitals ([10], p. 15). Boltzmann did, however, give us a fundamental quantum, the unit of information that Shannon would later call a "bit" [11]. Answering one yes/no question about a system reduces the entropy of the system, for the agent receiving the answer, by one bit. Over the course of the 20th century, quantum theory rendered objects more and more elusive, leading Bohr, Heisenberg, and others of a more empirical persuasion to focus ever harder on the role of measurement in answering questions. "It is no longer possible," wrote Bohr in 1937, "sharply to distinguish between the autonomous behavior of a physical object and its inevitable interaction with other bodies serving as measuring instruments, the direct consideration of which is excluded by the very nature of the concept of observation in itself" ([12], p. 290). Einstein and other realists were prompted, in turn, to ask whether Bohr and the other empiricists really thought the moon was not there if no one looked [13], and to construct ever more arcane "interpretations" of quantum measurement to preserve the observer-independence or "objectivity" of the objects being measured [14]. While the philosophical debate still rages on [15], the formalism of quantum theory has slowly become explicitly information-theoretic. Explaining his slogan, "it from bit," Wheeler wrote in 1990, "every it—every particle, every field of force, even the spacetime continuum itself—derives its function, its meaning, its very existence entirely—even if in some contexts indirectly—from the apparatus-elicited answers to yes or no questions, binary choices, bits" [16]. With this statement, Wheeler made Boltzmann's quantum of information the fundamental currency of physics.

1.2   Systems and Interfaces

If physics is about the answers to yes/no questions, who or what is asking them, and who or what is answering? Is "the observer" necessarily a human, or could it be any living system, or even any system whatsoever? Does the observer question a macroscopic apparatus, or does the observer employ an apparatus to question something else, an elementary particle perhaps, or a quantum field? What effect does this questioning have on the system questioned, or on the larger surrounding environment? Chap. 2 will address these questions directly; here we will focus on two even deeper questions. First, what distinguishes the question asker from the question answerer? And second, what enables the question to be asked, and the answer to be given? While such questions have long been posed by philosophers, physicists were only forced to face them in the 20th century. They were forced to do so, moreover, not by physics itself, but by technology.

When faced with something that behaves in an unexpected way, a natural inclination for at least a subset of humans—including those inclined toward science or engineering—is to take the thing apart and see how it works. This extends, however clumsily, even to living organisms, where putting the thing back together and having it still work is only possible in special cases. The first decisive, practical failure of this strategy occurred in the late 19th century, with the installation of the first undersea telegraphy cables. Problems somewhere on the bottom of the ocean could only be diagnosed, and if possible, worked around, by sending signals down the cable from one end and seeing what came out from the other end. Input/Output (I/O) analysis thus became part of the engineer's toolkit. By mid-century, two world wars and a huge advance in the complexity of electromechanical and chemical technologies later, the new discipline of cybernetics had formalized the idea of a "black box" that could only be investigated by I/O analysis [17]. A substantial movement within psychology had decided that human and animal behavior only needed to be studied by such analysis [18]. Biologists rejected this simplification, driving a wedge between biology and psychology that has still not been entirely removed.

The fundamental idea of the black box is the idea of an I/O interface through which interactions between the internals of the box and an external observer, experimenter, or analyst must flow. In the 1950s—prior to the era of microelectronics—the interface was envisioned as a "front panel" containing some finite number of dials or switches with which to affect the box's internal state and some finite number of meters or gauges that reported it, together with a "back panel" that had the power input to the box and the exhaust for waste heat. An experimenter could manipulate the inputs, including the power supply, and observe the outputs, including the exhaust, in an attempt to figure out what was inside the box and how whatever was inside worked. Such manipulations were not without risk; real-life "black boxes" recovered from an enemy, for example, could well be booby traps or bombs.

Edward Moore proved, in 1956, the fundamental theorem of black box analysis: that no finite sequence of manipulations of the inputs and observations of the outputs was sufficient to determine the "machine table"—an abstraction for the internal mechanism—of an arbitrary black box [19]. The results of an I/O analysis could not, in other words, ever be regarded as definitive. Moore's proof is deceptively simple: it considers a box that counts the manipulations, and after some finite number, changes its behavior. If the box has this structure, no sequence of I/O observations can prove that the next observation will not produce a radically different result. As Moore pointed out, this is no different from the situation faced by any empirical science: finite experiments can determine that a hypothesis is incorrect, but cannot prove that it is correct.

Moore's theorem for black boxes can be seen as the engineer's, or the experimentalist's, version of the notion of implementation independence that emerged from Church's [20] and Turing's [21] proofs—published months apart in 1936—that any procedure satisfying intuitive criteria for being a "computation" could be implemented by the lambda calculus or the Turing machine, respectively. These proofs, and their many subsequent elaborations, made it clear that one could not determine, by looking at the statement of a computational problem and its answer, how the answer had been computed. The vast array of programming languages and computational architectures available today is a consequence of implementation independence, as are virtual-machine architectures and application-programming interfaces, including user interfaces, that "hide the details" of how a computation is being implemented.

Both cybernetics and computer science, therefore, answer the question of what distinguishes the observer from the observed with the idea of a boundary that serves as an I/O interface. This boundary hides the details of what is happening inside the system being observed; it also, from the system's point of view, hides the details of what is happening inside the observer. The boundary is also, by its very definition, what enables questions to be asked and answered. This idea of a boundary that separates observer from observed, and that enables but also restricts the flow of information between them has been rediscovered and renamed over and over in the latter part of the 20th century. We will encounter it as the *Holographic Principle* of physics in Chap. 2, and as the idea of a *Markov Blanket* in Chap. 5.

1.3   Computation as Semantic Interpretation

The proofs by Church and Turing that effectively founded the discipline of computer science depended, as mentioned above, on an intuitive notion of "computation" as a finite, step-by-step procedure that produced an answer from a problem statement, each step of which could be checked for correctness. This intuitive notion is, effectively, the notion of a mathematical proof, and was regarded, prior to Church and Turing, as completely abstract—something happening in Plato's world of ideal forms or Descartes' disembodied mind. While both the lambda calculus and the Turing machine are also abstractions, the wartime computers that immediately followed were not. By 1945, the von Neumann architecture had been defined and implemented as the EDVAC [22], setting the stage for the rapid development of practical computer hardware and then software. "Computation" increasingly came to mean *physically implemented* computation. Of course, it had been physically implemented computation all along; only lingering Cartesian dualism had prevented this being recognized. McCulloch and Pitts' model of neural activity as computation [23] began the process of seeing the brain as a computer, and served as the basis for the development of artificial neural networks as machine learning systems in the 1980s [24], and hence for the deep neural nets, transformers, and diffusion models that dominate artificial intelligence today.

The McCulloch and Pitts model also, critically, introduced the idea that the measured behavior of a physical system could be described in computational terms. This idea is already implicit in the Turing machine, but seeing it requires reversing Turing's reasoning: seeing the "machine" as existing and operating—moving its head and writing and erasing symbols—before being thought of as a computer. It particularly inspired Wiener's development of cybernetics as an explicit practice of modeling physical systems in information-processing terms [25]. The general form of such models is shown in Fig. 1.1: given a physical system S that evolves through time t under the action of some propagator $\mathcal{P}(t)$, one specifies a mapping $\mathcal{M}$ and searches for a function f such that $f|_{t_i \to t_f} : \mathcal{M}(S(t_i)) \mapsto \mathcal{M}(S(t_f))$. This mapping $\mathcal{M}$ is an *interpretation* of the behavior of S as a computation of the function f during the behavioral epoch $t_i \to t_f$.

In line with Moore's theorem regarding interactions with black boxes, the key to any computational model is that the interpretation map $\mathcal{M}$ has finite resolution in time, and therefore finite resolution in representing the state of S [26]. This renders the model states $\mathcal{M}(S(t_i))$ and $\mathcal{M}(S(t_f))$ representable as finite bit strings, and the function f a computation on finite bit strings, i.e. a computation that can be implemented by a Turing machine. Finite model states $\mathcal{M}(S(t_i))$ and $\mathcal{M}(S(t_f))$ also render $\mathcal{M}$ interpretable as the output function of a black box, and hence interpretable as a measurement. This equivalence between computational interpretation and measurement has since become central to the physical definition of computation [27].

A correct model $\mathcal{M}$ of S is *semantic* in Tarski's sense: the system S actually behaves in the way described by $\mathcal{M}$ [28]. We can also say that S *instantiates* $\mathcal{M}$; this latter with the

Fig. 1.1 Interpreting the behavior of a physical system S between t_i and t_f as computing a function f. The map $\mathcal{M}$ is the interpretation function or computational model of S

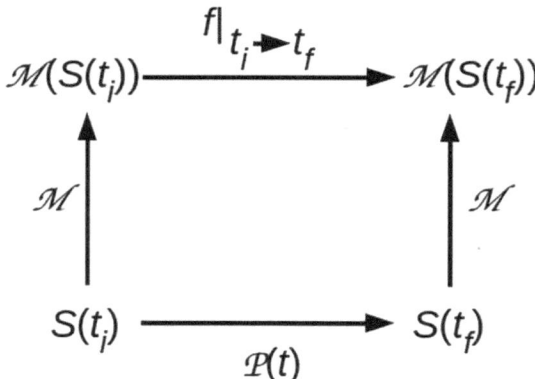

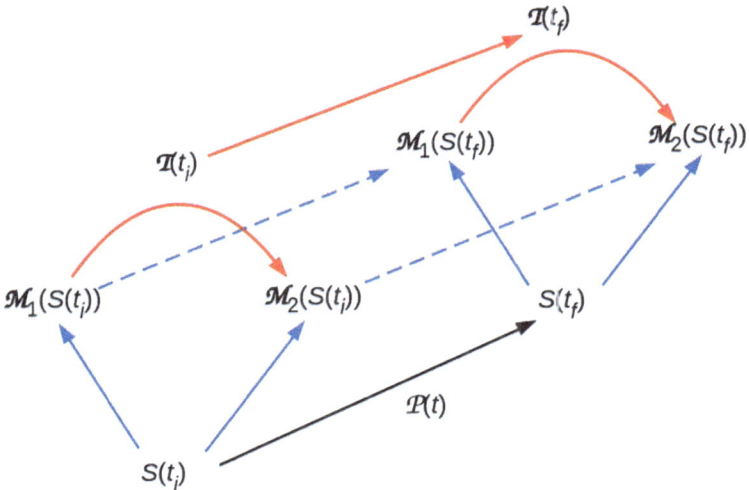

Fig. 1.2 Two different models $\mathcal{M}_1$ and $\mathcal{M}_2$ that interpret the behavior of a physical system S between t_i and t_f, with an inter-model map $\mathcal{T} : \mathcal{M}_1 \rightarrow \mathcal{M}_2$. Adapted from [29] Fig. 4; used with permission

understanding—reflecting implementation independence—that other systems may instantiate $\mathcal{M}$. It is also the case that multiple models can correctly describe the behavior of S; descriptions of the motion of an object in momentum versus position coordinates—Fourier analysis—provides a case in point. Hence in general we have situations of the form of Fig. 1.2, and can ask how the alternative models $\mathcal{M}_1$ and $\mathcal{M}_2$ relate, i.e. we can ask what the map $\mathcal{T} : \mathcal{M}_1 \rightarrow \mathcal{M}_2$ looks like. We can, in particular, ask whether $\mathcal{T}$ is time-invariant, assuring that $\mathcal{M}_1$ and $\mathcal{M}_2$ maintain the same inter-model semantic relationship. If $\mathcal{M}_1$ and $\mathcal{M}_2$ are descriptions of S using different coordinates as above, this is the question of whether the mapping between coordinate systems, e.g. the mapping $x \mapsto (md/dt)(x)$, is well-defined.

If S is a computer and $\mathcal{M}_1$ and $\mathcal{M}_2$ are models constructed using two distinct programming languages, Fig. 1.2 describes the relationship between two programs running simultaneously on S. This situation is, of course, familiar, and forms the basis for concurrency and virtual machine hierarchies [30]. It is also commonplace in biology, where Bongard and Levin have dubbed it "polycomputing" [31]. Time invariance of $\mathcal{T}$ is required, in the first instance, for the relationship between programming languages to be well-defined. It is required in the second instance for the relationship between alternative descriptions of a biological system to be well-defined. We will see in Chap. 2 below that this requirement follows from the use of the tensor product to decompose state spaces, and hence from the representation of state spaces by vector spaces.

1.4   Scale-Free Versus Reductionist-Emergentist Theories

As noted earlier, it is a natural approach to understanding the behavior of a complex system to take the system apart and figure out what its parts are doing. If the behaviors of the parts, when taken together, fully explain the behavior of the composite system, one can construct a successful "reductionist" theory of system-level behavior. Prior to the development of microprocessors, this strategy worked for electromechanical devices, rendering "reverse engineering" an effective learning strategy for people who would later practice forward engineering as careers. It also worked, famously, from the scale of small-molecule chemistry downward to that of subatomic particles. It has not, however, worked for any living systems, or even for the behavior of large molecules in vivo. When some system-level behaviors remain unexplained by a reductionist theory, it is commonplace to call them "emergent" and look for a special, scale-specific explanation that involves novel concepts that do not apply at the "microscale" of the parts.

Reductionist theories go hand-in-hand with the idea that *explaining* a phenomenon is deducing its occurrence from some set of laws of Nature together with some set of boundary conditions [32]. The fundamental laws of Nature, in this "nomological deductivist" framework, are stated at some microscale, e.g. the scale of the Standard Model of particle physics, so explaining macroscale phenomena requires recourse to some "bridge laws" that connect macro- to microscale behavior. Examples of these that are typically encountered range from informal definitions like "heat is molecular motion" to structure theories that specify which nuclei, which atoms, or which molecules can remain stable, and for how long. In the language of Fig. 1.2, if $\mathcal{M}_1$ and $\mathcal{M}_2$ are stated at different scales, $\mathcal{T}$ must be a bridge law. Clearly to be a law, $\mathcal{T}$ must be time-invariant.

In the absence of bridge laws, macroscale phenomena are emergent, and have no nomological-deductivist explanation. It is commonplace to distinguish "weak" from "strong" senses of emergence; a behavior or property of S is said to be weakly emergent at a (relative) macroscale if it is reproduced by a simulation of the collective behavior of S's (relative) microscale components [33], and is strongly emergent if such a simulation fails to reproduce the behavior or property. This distinction clearly depends on the quality of both the microscale theory available and the simulation, including choices of initial and boundary conditions. A weakly emergent behavior or property can be viewed as explained by the simulation within the nomological-deductivist framework if simulation is accepted as a kind of automated deduction that may be intractable for humans, but tractable by computers. The question of whether any behaviors or properties of any systems could be strongly emergent in principle, and hence resistant to any kind of analysis and reconstruction, is vigorously debated by philosophers [34].

There is, however, a natural alternative to the dichotomy between reductionist and emergentist approaches to explanation: the idea that behavior, and hence its explanation, could be *scale free*. A behavior is scale free if it occurs, in the same form, at every scale, and its explanation is scale free if it can be applied, in the same form, at every scale. Natural selection in biology provides an example: it acts at every scale from the macromolecular to the ecological, and has the same formal dynamics of variation followed by differential reproduction at every scale. This dynamics can be abstracted out of biology, and applied formally to optimization problems in general [35]. Black-box cybernetics is similarly scale-free, as is the theory of computing. The latter provides the natural interpretation of Fig. 1.2 for scale-free theories: if $\mathcal{M}_1$ and $\mathcal{M}_2$ are stated at different scales, $\mathcal{T}$ is the implementation relation. What is of interest in a scale-free theory is, therefore, how the formal dynamics at one scale is implemented by the same formal dynamics at a smaller scale. The theory that we will develop in the remainder of this book is scale-free in this sense.

References

1. Clausius, R. The Mechanical Theory of Heat–with Its Applications to the Steam Engine and to Physical Properties of Bodies; John van Voorst: London, UK, 1867.
2. Boltzmann, L. On the relationship between the second fundamental theorem of the mechanical theory of heat and probability calculations regarding the conditions for thermal equilibrium. Sitz. Kaiserlichen Akad. Wissenschaften Mathematisch-Naturwissen Classe Abt. II 1877, LXXVI, 373–435. (Translated and annotated by Sharp, K.; Matschinsky, F. Entropy 2015, 17, 1971–2009).
3. Boltzmann, L. *Lectures on Gas Theory*. Dover, New York NY, 1995.
4. Laplace P-S (1814) Essai philosophique sur les probabilités. Paris: Courcier. (Available as Philosophical Essay on Probabilities, Dover, New York, 1996)
5. Landauer, R. Irreversibility and heat generation in the computing process. IBM J. Res. Dev. 1961, 5, 183-195.
6. Bennett, C.H. The thermodynamics of computation. Int. J. Theor. Phys. 1982, 21, 905-940.
7. Landauer, R. Information is a physical entity. Physica A 1999, 263, 63-67.
8. Parrondo, J.M.R.; Horowitz, J.M.; Sagawa, T. Thermodynamics of information. Nat. Phys. 2015, 11, 131-139.
9. Wick, G. C. (1954). Properties of Bethe-Salpeter Wave Functions. Physical Review 96, 1124-1134.
10. Bohr N (1913) On the constitution of atoms and molecules. Philos Mag 26, 1-25.
11. Shannon, C. E. A mathematical theory of communication. Bell System Technical Journal. 27, 379-423.
12. Bohr, N. Causality and complementarity. Philos. Sci. 1937, 4, 289-298.
13. Pais, A. (1979) Eistein and the quantum theory. Rev. Mod. Phys. 51, 863-914.
14. Landsman, N.P. Between classical and quantum. In Handbook of the Philosophy of Science: Philosophy of Physics; Butterfield, J., Earman, J., Eds.; Elsevier: Amsterdam, The Netherlands, 2007; pp. 417-553.
15. Cabello, A. Interpretations of quantum theory: A map of madness. In: Lombardi, O., Fortin, S., Holik, S., López, C. (Eds.) *What is Quantum Information?* Cambridge University Press, Cambridge, UK, 2017, pp. 138-140.
16. Wheeler, J.A. Information, physics, quantum: The search for links. In Complexity, Entropy, and the Physics of Information; Zurek, W.H., Ed.; Westview: Boulder, CO, USA, 1990; pp. 3–28.
17. Ashby, W.R. Introduction to Cybernetics; Chapman and Hall: London, UK, 1956.
18. Watson, J.B. (1913). Psychology as the behaviorist views it. Psychological Review, 20, 158-177.

19. Moore, E.F. Gedankenexperiments on sequential machines. In Autonoma Studies; Shannon, C.W., McCarthy, J., Eds.; Princeton University Press: Princeton, NJ, USA, 1956; pp. 129-155.

20. Church, A. [1936] An unsolvable problem of elementary number theory, Am. J. Math. 58, 345-363.

21. Turing, A. M. [1936] On computable numbers, with an application to the Entscheidungsproblem, Proc. Lond. Math. Soc. (2) 42, 230-265.

22. von Neumann, J. (1945) First draft of a report on the EDVAC. Moore School of Electrical Engineering, University of Pennsylviania. Reprinted as *IEEE Annals of the History of Computing* 15(4), 1993, pp. 29-43.

23. McCulloch, W. S., Pitts, W. (1943) A logical calculus of the ideas immanent in nervous activity. Bulletin of Mathematical Biophysics. 5(4), 115-133.

24. Rumelhart, D. E., Hinton, G. E., Williams, R. J. (1986) Learning representations by back-propagating errors. Nature 323, 533-536.

25. Weiner, N. (1948) Cybernetics: Or Control and Communication in the Animal and the Machine. MIT Press, Cambridge, MA, USA.

26. Fields, C. Consequences of nonclassical measurement for the algorithmic description of continuous dynamical systems. J. Expt. Theor. Artif. Intell. 1989, 1, 171-178.

27. Horsman, C.; Stepney, S.; Wagner, R.C.; Kendon, V. When does a physical system compute? Proc. R. Soc. A 2014, 470, 20140182.

28. Tarski, A. (1944) The semantic concept of truth and the foundations of semantics. Philosophy and Phenomenological Research 4, 341-375.

29. Fields, C. and Levin, M. (2025). Thoughts and thinkers: On the complementarity between objects and processes. Physics of Life Reviews 52: 256–273

30. Smith, J. E., Nair, R. (2005) The architecture of virtual machines. IEEE Computer 38(5), 32-38.

31. Bongard, J., Levin, M. (2023) There's plenty of room right here: Biological systems as evolved, overloaded, multi-scale machines. Biomimetics 8, 110.

32. Hempel, C., Oppenheim, P. (1948) Studies in the logic of explanation. Philosophy of Science, 15, 135-175.

33. Bedau, M. A. (1997) Weak emergence. In: J. Tomberlin (ed.) Philosophical Perspectives, Vol. 11, Mind, Causation, and World. Blackwell, Malden, MA, USA, pp. 375-399.

34. Clayton, P., Davies, P. C. W., Eds. (2006) The Re-emergence of Emergence. Oxford University Press, Oxford, UK.

35. Holland, J. (1992) Adaptation in Natural and Artificial Systems. MIT Press, Cambridge, MA, USA.

Generic Quantum Systems

2

We will be concerned, in this chapter, with the idea of a *system* S that is embedded in and interacts with another system, conventionally called the *environment* E_S of S, which is also a system. The fact that S interacts with E_S—we will often drop the 'S' subscript and just call this 'E' when the context is clear—renders S an *open system*; E, by virtue of interacting with S, is also an open system. We will sometimes, beginning in Sect. 2.3 below, describe S as an "observer" of E and thus develop a generic theory of S "preparing" a state of E and then making a "measurement" of that prepared state. We will state and prove a number of "no-go" theorems that restrict the information that S can obtain by making such measurements.

2.1   Closed Versus Open Systems

2.1.1   The "Universe" U

Let U be a system that can properly be called a *universe*, i.e. U includes "everything" or at least "everything of interest." If U contains everything, it can have no environment; hence U is *isolated* or *closed*. We can, in practice, define U as the unique system that has no environment, as there is no point in talking about multiple, mutually isolated systems.

We assume that U can be characterized by, or simply comprises, some finite number n of *degrees of freedom* or state variables, each of which has some finite number m of possible values. We can, therefore, represent a state $|U\rangle$ of U—we will use the Dirac notation for states—by specifying nm values. To acknowledge that U is "large" we allow nm to be arbitrarily large; the assumption that nm can be large but remains finite is made for convenience, to avoid having to later assume that specific physical quantities remain finite. To specify states of U, it is convenient to choose a basis $|i\rangle = |i_1, i_2, \ldots i_{nm}\rangle$ so that an eigenvector is a

© The Author(s), under exclusive license to Springer Nature Switzerland AG 2026   11
C. Fields and J. Glazebrook, *Distributed Information and Computation in Generic Quantum Systems*, Synthesis Lectures on Engineering, Science, and Technology, https://doi.org/10.1007/978-3-031-97263-8_2

binary string, with $i_j = 1$ if and only if the state represented by the eigenvector has the jth value. A complex Hilbert space with this basis is a *qubit*, or quantum bit, space $\mathcal{H}_U = q^{nm}$ of dimension 2^{nm}. The basis $|i\rangle$ is conventionally called the *computational basis* for $\mathcal{H}_U$ [1].

Let t_U be a real parameter interpreted as an abstract "time" through which the state $|U(t_U)\rangle$ evolves, and let $\mathcal{P}_U(t_U) = \exp((-\iota/\hbar)H_U(t_U))$, with $\mathcal{P}_U$ unitary, i.e. t_U-symmetric, be the time-propagator for $|U(t_U)\rangle$. In this case $H_U : \mathcal{H}_U \rightarrow \mathcal{H}_U$ is a Hermitian operator, the *Hamiltonian* for U, the eigenvalues of which are the possible values of the total energy $\mathbf{E}_U$ of U. The assumption that nm remains finite assures that $\mathbf{E}_U$ remains finite. The assumption that U is closed assures that U contains no energy sources or sinks, so provided H_U is time-invariant, we can choose a zero point of energy, i.e. stipulate that $\mathbf{E}_U = 0$.

We can now ask a more philosophical question: what is U, really? We will see below—and indeed will emphasize where appropriate—that it makes not a shred of difference whether we think of a "thing" U evolving through "time" t_U or a "process" $\mathcal{P}_U$ defined as a function of t_U. These ways of thinking correspond to the "Schrödinger picture" that focuses on systems and their states and the "Heisenberg picture" that focuses on operators, respectively. We will in what follows move between these "pictures" whenever one is more convenient than the other.

2.1.2 Decompositional Equivalence

Using the vector space tensor product operator $\otimes$, we can decompose $\mathcal{H}_U$ as $\mathcal{H}_U = \mathcal{H}_S \otimes \mathcal{H}_{E_S}$ for any pair of systems S, E_S that jointly compose U. In this case, the state $|U\rangle$ can be written as the joint state $|U\rangle = |SE_S\rangle$. The Hamiltonian H_U decomposes additively as $H_U = H_S + H_{E_S} + H_{SE_S}$, where H_S and H_{E_S} are the self-interactions of S and E_S, respectively, and H_{SE_S} is the interaction between S and E_S.

These simple, mathematical facts have an important consequence:

Where the boundary between S and E_S is drawn makes no difference to the physics of U.

Where the boundary between S and E_S is drawn obviously makes a difference to the physics of S and E_S. But differences in how S and E_S are defined, and hence differences in how H_S, H_{E_S}, and H_{SE_S} are defined make no difference to the physics of U. We will emphasize these facts over and over in what follows, in places where their consequences may be counter-intuitive or even jarring.

2.1.3 Separability of S from E_S

Given a basis for $\mathcal{H}_U$—e.g. the computational basis $|i\rangle$—and a decomposition $U = SE_S$, the joint state $|U\rangle = |SE_S\rangle$ is *separable* if and only if $|SE_S\rangle = |S\rangle|E_S\rangle$, and is *entangled* otherwise. Whether a joint state is separable depends sensitively on how it is described [2]; for example, the Bell states $(|\uparrow\rangle \pm |\downarrow\rangle)/\sqrt{2}$ are maximally entangled when described in the $(|\uparrow\rangle, |\downarrow\rangle)$ basis, but $|\uparrow\rangle$ and $|\downarrow\rangle$ are maximally entangled if $(|\uparrow\rangle \pm |\downarrow\rangle)/\sqrt{2}$ are chosen as the basis. Nothing *physical* restricts the choice of basis for U; it is purely a theoretical stipulation. Hence we can state:

Separability of quantum states, and hence separability of quantum systems, is stipulated, not observed.

Intuitively, S and E_S are separable if, but only if, $|S\rangle$ depends on $|E_S\rangle$ only causally; in this case, we can regard S and E_S as conditionally statistically independent. Failures of conditional statistical independence manifest as failures of the Kolmogorov axioms for the joint probability distribution $P(|S\rangle, |E_S\rangle)$ and hence as violations of Bell's inequality [3], which can be stated in its conventional Clauser-Horne-Shimony-Holt (CHSH) form as:

$$EXP = |\mathbf{e}(S_1, E_1) + \mathbf{e}(S_1, E_2) + \mathbf{e}(S_2, E_1) - \mathbf{e}(S_2, E_2)| \leq 2 \qquad (2.1)$$

where S_i and E_j, $i, j \in \{1, 2\}$ are mutually independent observables for S and E_S, respectively, that can have measured values of $+1$ or -1, and $\mathbf{e}(x, y)$ indicates the expectation value of a joint measurement of x and y [4].

In practice, systems may be approximately separable, i.e. only minimally entangled, and hence interact approximately causally. Given a decomposition $U = SE_S$ and a basis $|i\rangle$ for U, the entanglement entropy $\mathcal{S}(\rho_S) = -\text{Tr}(\rho_S \ln \rho_S)$, where the density $\rho_S = |S\rangle\langle S| = \text{Tr}_{S_E}(|SE_S\rangle\langle SE_S|)$ and Tr is the trace operator. As $\mathcal{S}$ measures the entanglement of $|SE_S\rangle$, $\mathcal{S}(\rho_S) = \mathcal{S}(\rho_{E_S})$. If $|SE_S\rangle$ is separable, $\mathcal{S}(\rho_S) = 0$; if $|SE_S\rangle$ is maximally entangled, $\mathcal{S}(\rho_S) = \min(\dim(\mathcal{H}_S), \dim(\mathcal{H}_{E_S}))$. Hence empirical measurements of CHSH inequality violations are empirical demonstrations of the failure of separability. In line with the statement above, such empirical measurements are empirical *with respect to* a stipulated choice of basis. This bears emphasis:

There is no "objective" observer-independent entanglement.

Statements about entanglement or separability are always with respect to some stipulated choice of basis, whether this is stated explicitly or not. We will see in Sect. 2.3 below that choosing a basis is equivalent to choosing a reference frame, or a method of measurement. Hence whether a system is entangled or separable depends on how its state is measured.

Intuitively, $\mathcal{S}(\rho_S) \to 0$ as the interaction $H_{SE_S} \to 0$, with $H_{SE_S} = 0$ indicating isolation of S from E_S, i.e. S ceasing to be an open system. Hence approximate separability corresponds to a weak interaction limit in which H_{SE_S} minimally perturbs $|S\rangle$. If we think of the situation semi-classically, considering S to be decomposable as $S = S_1 S_2$ such that $|S_1 S_2\rangle = |S_1\rangle |S_2\rangle$, i.e. if S is itself separable, we can imagine that H_{SE_S} acts "directly" only on S_1, after which S_1 acts causally on S_2. In this case, S_2 can be regarded as an "internal" component of S that is "shielded" from E_S by S_1. This semi-classical picture will prove useful when modeling S as a semi-classical computer in Chap. 4 below.

2.2   Information Flow and the Holographic Principle

If S and E_S are at least approximately separable, it makes sense to talk about S sending classical information to, or "writing" a sequence of classical bits on E_S. This intuition will now be developed formally. To simplify the notation, we will drop the subscript on E_S, simply writing $U = SE$, H_{SE}, and so on whenever the context is clear.

2.2.1   Defining H_{SE}

Recall Landauer's Principle, $d\mathbf{E} = \beta k_B T$ for irreversibly writing one bit of information, with $\beta \geq \ln 2$ an inverse measure of thermodynamic efficiency, from Chap. 1, Sect. 1.1. If S irreversibly writes N bits on E with a constant efficiency β_S, S's energetic cost is $d\mathbf{E}_S = N\beta_S k_B T_S$. Since U is closed, this energy must come from E. If E in turn irreversibly writes N bits on S with a constant efficiency β_E, E's energetic cost is $d\mathbf{E}_E = N\beta_E k_B T_E$, and this energy must come from S. If this exchange of energy is the only interaction between S and E, the Hamiltonian H_{SE} being well-defined requires that it has the single eigenvalue $d\mathbf{E}_S = d\mathbf{E}_E$, in which case we have $\beta_S T_S = \beta_E T_E$.

It is straightforward to construct an instantaneous (time-independent) interaction Hamiltonian with this eigenvalue. Letting M_i^k, $k = S$ or E, be an operator that writes, or dually, reads one bit, we can write:

$$H_{SE} = \beta_k k_B T_k \sum_{i=1}^{N} M_i^k \qquad (2.2)$$

for $k = S$ or E. For convenience, we assume the M_i^k are orthonormal with eigenvalues of either $+1$ or -1. Writing one classical bit can be physically implemented by preparing one qubit in either $|\uparrow\rangle$ or $|\downarrow\rangle$ with respect to some local z axis. Hence we can visualize the action of H_{SE} as in Fig. 2.1, noting that in principle a distinct local z axis may be chosen for each qubit q_i by each of S and E. The collection of mutually-independent, non-interacting

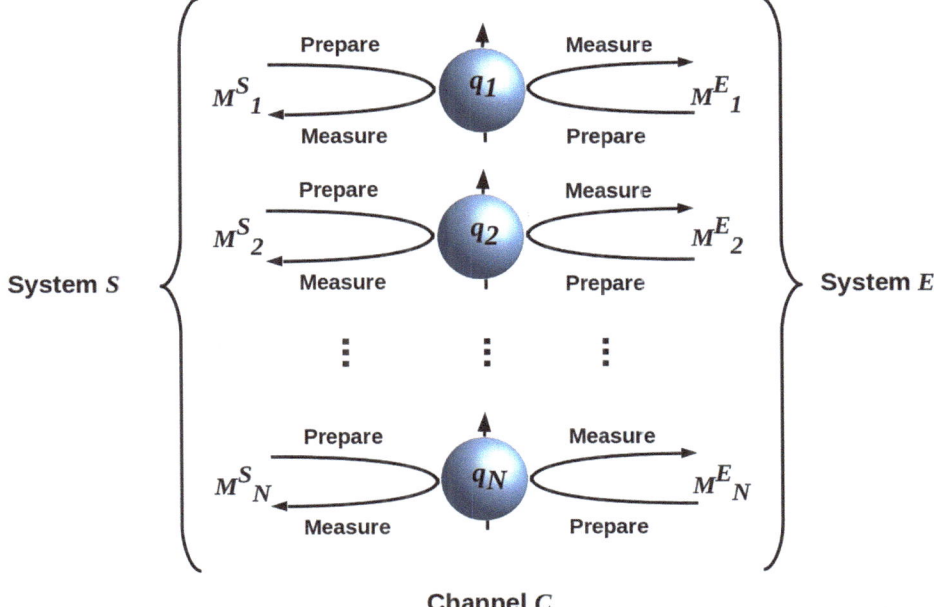

Fig. 2.1 Interpreting H_{SE} as a sum of single-qubit operators. The qubits q_i compose the informational boundary between S and E, which the Holographic Principle associates with the decompositional boundary between S and E. Adapted from [5] Fig. 1, CC-BY license

qubits $q_1, q_2, \ldots q_N$ constitute, in this implementation, an *informational boundary* between S and E.

The operator decomposition $H_U = H_S + H_E + H_{SE}$ is induced by the Hilbert space decomposition $\mathcal{H}_U = \mathcal{H}_S \otimes \mathcal{H}_E$. The interaction H_{SE} can, therefore, be regarded as *defined on* the decompositional boundary $\mathscr{B}$ that is implied by $\mathcal{H}_U = \mathcal{H}_S \otimes \mathcal{H}_E$. It is this boundary $\mathscr{B}$ that "separates" S from E whenever $|SE\rangle$ is separable.

2.2.2   The Holographic Principle

Recall from Chap. 1 that when Clausius introduced the notion of the classical entropy, $S(X)$ of a system X, this entropy was a pure abstraction. It remained so well into the 20th century, with John von Neumann famously advising Claude Shannon to refer to his new information measure as "entropy" because "no one knows what entropy really is" [6]. The idea that entropy could only be measured at the interface between two systems, while implicit in both Shannon's Information Theory and the black box model of cybernetics, was only made explicit in 1974, when Jakob Bekenstein showed that the classical entropy of a black hole (BH) was $\mathbf{S}(BH) = A_{BH}/4$, where A_{BH} is the area of the "stretched" BH horizon in Planck units, i.e. units in which the numerical values of $\hbar$, c, and k_B are all unity [7]. This association between entropy and area was generalized by Gerardus 't Hooft, who stated the Holographic Principle (HP) in 1993: "Given any closed surface, we can represent all that happens inside it by degrees of freedom on this surface itself." Specifically, the classical entropy $\mathbf{S}(X)$ of any closed system X is $\mathbf{S}(X) \le A_X/4$, again in Planck units [8]. A specific mechanism for encoding $\mathbf{S}(X)$ on the surface of X was given by Leonard Susskind in 1995: a photon leaving X could encode 1 bit on the surface in an area no smaller than $4l_P^2$ [9]. The HP was generalized to surfaces that are not closed, and given a covariant formulation in terms of light sheets associated by X, by Rafael Bousso [10]. The HP was, however, still regarded as something of an oddity of quantum cosmology, with Bousso remarking that the HP remained:

> ... an apparent law of physics that stands by itself, both uncontradicted and unexplained by existing theories ... [that] may still prove incorrect or merely accidental, signifying no deeper origin. [10], p. 826.

When one considers the intersystem boundary $\mathscr{B}$ shown in Fig. 2.1, however, it becomes obvious that the HP applies to any boundary between separable systems [11]. In this generic setting, the classical entropy $\mathbf{S}(\mathscr{B})$ encoded on $\mathscr{B}$ is just Nk_B, with N the number of orthogonal single-bit operators M_i^k in the expansion of the Hamiltonian H_{SE} defined at $\mathscr{B}$. The interaction H_{SE} transfers $\Delta E_k = \mathbf{S}(\mathscr{B})\beta_k T_k$ to system k, with differences in ΔE_k between systems due to differences in thermodynamic efficiency, or equivalently, in effective T_k.

We can, therefore, view the HP as enforcing a fundamental informational symmetry on generic physical interactions: in any interaction H_{SE} between finite, separable systems S and E, the information flow from S to E equals the information flow from E to S. Both flows are given by the classical entropy $\mathbf{S}(\mathscr{B})$ of the intersystem boundary $\mathscr{B}$. The HP is, in other words, a local statement of the Principle of Unitarity, which enforces information symmetry—and hence reversibility or time symmetry—globally. It is a quantum-theoretic analog of Newton's Third Law [12].

2.3 "The Observer" as a Quantum System

It is commonplace in discussions of quantum theory to see arbitrary quantum systems characterized as "observers"; Maximilian Schlosshauer [13], for example, states in his textbook that "We simply treat the observer as a quantum system interacting with the observed system" (p. 361). The canonical tasks of observers are preparation and measurement of quantum states, which are dual tasks that employ the same operators [14]. That arbitrary quantum systems can prepare and measure quantum states is made explicit by the expansion of H_{SE} into the operators M_i^k in Eq. (2.2) and is illustrated in Fig. 2.1. From a formal perspective, in other words, the characterization of arbitrary quantum systems as "observers" follows immediately from the HP.

Box 2.1: Quantum theory is a theory of measurement

An enormous literature is devoted to the *quantum measurement problem*, the problem of understanding how quantum states, e.g. the state $(|alive\rangle + |dead\rangle)/\sqrt{2}$ of Schrödinger's infamous cat, can produce classical states when measured. Both physicists and philosophers have tried to resolve this problem, proposing either mathematical modifications or metaphysical interpretations of quantum theory to explain it; see [15] for a particularly dispassionate review. Not all physicists take this problem seriously; Adán Cabello subtitled a recent survey of the variety of interpretations a "map of madness" [16], while Chris Fuchs recalls Hideo Mabuchi saying that "the quantum measurement problem refers to a set of people" [17], p. 13. David Mermin has recently argued that "there is no quantum measurement problem" [18]; the "problem" so named is rather an artifact of a mistaken assumption that measurement results should be "objective" in some principled sense (see Mermin's excellent [19] for a deep discussion of this assumption).

Quantum theory as presented here is, as Fig. 2.1 makes clear, itself a theory of measurement. It is, moreover, a theory of *personal* measurement as Mermin demands—only S can receive S's observational outcomes, because only S is measuring S's environment. The outcomes that S obtains are, moreover, in principle dependent on *all* of S's environment, via the internal Hamiltonian H_E. Measurement outcomes are classical because the eigenvalues of Hermitian operators are real numbers, but there is no reason to regard S or any other *system* as classical. Indeed the theory of measurement expressed by Fig. 2.1 is consistent with a purely-quantum universe in which classical information *only* exists as a momentary encoding on an ancillary holographic screen [12].

Quantum theory is, therefore, not a theory about things, but a theory about interactions, all of which can be considered preparation-and-measurement interactions. As such, it is completely scale-free, applying in exactly the same way to all systems from the Planck scale to the scale of the entire universe. The theory needs no interpretation; indeed, it tells us how "interpretation" or indeed, semantics of any kind, can be possible.

When the system S is viewed as an observer, it becomes immediately obvious that the system E can also be viewed as an observer. We can, in other words, view any interaction H_{SE} between finite, separable systems as an exchange of information—a "conversation"—between the two observers S and E. They exchange information by alternately preparing and measuring the qubits on $\mathscr{B}$. Preparing a qubit in some state is acting on it with a local z-spin operator σ_z. This operation requires a reference frame that defines a local z axis. Hence observers must implement reference frames.

2.4   Quantum Reference Frames (QRFs)

In classical physics, reference frames are abstractions; a Cartesian coordinate system, for example, is just a piece of mathematical formalism. If Alice wants to replicate Bob's experiment, she asks Bob for a formal description of his reference frame, and then implements it in her laboratory. The Methods sections of experimental papers, with their citations of equipment manufacturers and reagent suppliers, are considered sufficient to specify the reference frames employed, with private communications between researchers as a backup in cases of ambiguity or apparent differences in function.

2.4.1   QRFs as Physical Systems

Quantum theory challenges this classical practice by noting that reference frames must be physically implemented to be useful, that they are therefore *quantum* reference frames (QRFs), and that quantum systems are generically not amenable to finite description [20]. The "nonfungibility" of quantum information means that QRFs cannot be replicated from descriptions—indeed, to do so would violate the no-cloning theorem [21]—and hence must be physically transferred from one observer to another. Appeals to calibration standards, in this case, lead to an infinite regress. Even when pervasive QRFs such as the Earth's gravitational field are employed, subtle differences between locations remain, in principle, not fully characterizable.

Once reference frames are conceptualized as physical systems, and therefore as QRFs, it becomes evident that observers cannot rely exclusively on external QRFs, but must also implement them internally [22]. In Fig. 2.1, for example, S cannot rely on a QRF embedded in E to prepare or measure qubits on $\mathscr{B}$, as the preparation and measurement of qubits on $\mathscr{B}$ is required to extract information from E. Any such S must, instead, encode a local z axis—in principle, one for each qubit—with which to conduct preparation and measurement as operations on $\mathscr{B}$. Indeed, choosing the local z axis for each qubit is just choosing the Hilbert-space basis in which to express the operators M_i^S.

The QRFs implemented by S are components of the physical system S and are, therefore, implemented by the physical dynamics specified by the Hamiltonian H_S. "Choosing a local z-axis for the ith qubit" is, in this case, an action of H_S. This action of H_S cannot be determined by the action of H_E without violating separability [12]; it is therefore "free" in the sense of "free choice" required by the Conway-Kochen theorems [23, 24], and regularly assumed in discussions of Bell/EPR [25] or delayed-choice [26] experiments.

Box 2.2: Where's the physics?

The Hamiltonian H_{SE} defined by Eq. (2.2) is time-independent. When we talk about "choosing a local z-axis" or "an action of H_S," however, we are talking about something happening, so are talking about time. The only time coordinate that has been defined is the "universal" time t_U that parameterizes the propagator $\mathcal{P}_U(t_U) = \exp((-\iota/\hbar)H_U(t_U))$. As U is isolated by definition, t_U is not an observable time. It provides, however, a natural time parameter for H_{SE} that will be used in Sect. 2.5.2 below to define a system-relative, observable time.

If we think of $\mathscr{B}$ as an array of qubits as in Fig. 2.1, it is natural to think of the operator M_i^S acting on a qubit q_i on $\mathscr{B}$ as an instance of the z-spin operator σ_z. We then need to specify a local z-axis with respect to which the local σ_z acts. We can then write:

$$M_i^k = \exp(\iota \phi_i^k(t_U))\sigma_{\phi_i^k} \qquad (2.2.1)$$

The phase angle $\phi_i^k(t_U)$ specifies the local z-axis associated with M_i^k. As S and E have free choice of QRFs, or of Hilbert-space bases for specifying the M_i^k, there is no requirement that $\phi_i^S = \phi_i^E$, and they can be assumed generically to be distinct.

Making $\phi_i^k(t_U)$ a QRF requires a physical implementation, e.g. by a qubit, or by multiple qubits that implement a quantum error-correcting code (QECC) [27]. These qubit degrees of freedom must be degrees of freedom of the system k; hence we can write, for some sufficiently small dt_U:

$$\mathcal{P}_k|_{t_U \to t_U + dt_U} : \phi_i^k(t_U) \mapsto \phi_i^k(t_U + dt_U) \qquad (2.2.2)$$

where $\mathcal{P}_k(t_U)$ is the propagator for system k. This propagator can be considered unitary only if k is isolated, i.e. we can write $\mathcal{P}_k(t_U) = \exp((-\iota/\hbar)H_k(t_U))$ only during intervals dt_U smaller than the time required for k to prepare or measure qubits on $\mathscr{B}$. We can interpret dt_U, in this limit, as the time required for k to "choose the QRFs" ϕ_i^k for each cycle of preparation and measurement operations on $\mathscr{B}$.

2.4.2   Breaking Permutation Symmetry on $\mathscr{B}$

As described thus far, the boundary $\mathscr{B}$ is symmetric under permutations of the qubits q_i; any such permutation just relabels the basis vectors of the Hilbert space $\mathcal{H}_{\mathscr{B}}$. The notion of S or E preparing or measuring states of $\mathscr{B}$ does not, therefore, yet have much content, as nothing has been said about how either S or E distinguishes one state of $\mathscr{B}$ from another. Distinguishing between states of $\mathscr{B}$ requires breaking the permutation symmetry; hence it requires stipulating that the internal Hamiltonians H_S and H_E act in distinct ways on distinct qubits q_i on $\mathscr{B}$. Recall from Chap. 1, Sect. 1.3 that a system S can be regarded as *computing* if an interpretation map $\mathcal{M}$ and a function f can be found such that S's state transitions commute with f, i.e. such that $f\mathcal{M} = \mathcal{M}\mathcal{P}_S$. Computation, clearly, involves *doing something interesting* with inputs to produce outputs. We can, therefore, say:

An *interesting* system is one that computes some interesting function.

We will be interested, in the remainder of this book, in interesting systems, and will delve in great detail into what makes them interesting.

Part of the idea of computation is that when a function f is computed, it stays computed. The *process* of computation may be reversible, but the *result* of a computation is not. A result can be written down irreversibly, and counted on not to change. It does not need to be computed again, and if it is, the same result will be obtained. It is these features of the idea of computation that enable it to be thought of as a purely abstract process, one that is entirely independent of contingent facts.

This irreversibility built into the idea of computation gives every physically-implemented computation a minimal, implementation-independent energetic cost: the cost of writing down the result. Landauer's principle sets the value of this minimal cost to $\ln 2 k_B T$ per bit [28]. This cost is built into Eq. (2.2), scaled by the system-specific efficiency factor $\beta_k \geq \ln 2$. As noted above, this energy must come from E. It must, therefore, be transmitted from E to S via qubits on $\mathcal{B}$.

We can now make an observation: the *quantity* of bits "burned as fuel" to record results of computations matters, but the numerical values of the bits burned as fuel do not. Bits burned as fuel are not, therefore, informative as inputs to a computation, and they cannot encode outputs. The input and output devices of a computer, for example, are distinct from its power supply. This distinction between informative and fuel bits breaks the symmetry of the boundary $\mathcal{B}$ in any system S that performs computation, i.e. in any interesting system [5, 29]. It defines two distinct, disjoint sets of qubits, which we will refer to as *sectors* of $\mathcal{B}$: the *informative sector* χ that encodes computational inputs and outputs, and the *thermodynamic sector* Θ that provides the needed power. It similarly breaks the symmetry of H_S, dividing it into a computational component and a thermodynamic (or power supply) component. The thermodynamic efficiencies of these components—or equivalently, their operating temperatures—must be such that thermodynamic free energy (TFE) flow through the thermodynamic component is sufficient to meet the power requirements of the computational component, i.e.

$$N_\Theta \beta_\Theta T_\Theta \geq N_\chi \beta_\chi T_\chi \tag{2.3}$$

where N_Θ and N_χ are the respective numbers of qubits on $\mathcal{B}$, for the thermodynamic and information sectors of any system S. This symmetry breaking is illustrated in Fig. 2.2.

The computational and thermodynamic components Q_χ and Q_Θ of H_S can, without loss of generality, be represented as QRFs [31]; see Chap. 7 for discussion and proof. The thermodynamic QRF Q_Θ, in particular, sets the energy scale for S. The two-QRF architecture shown in Fig. 2.2 is the minimal system architecture that will be of interest for

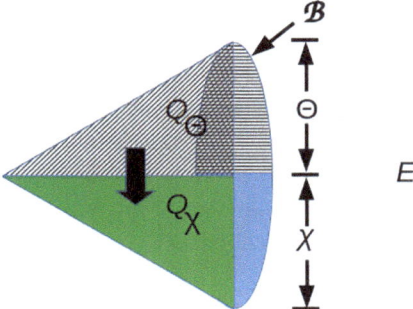

Fig. 2.2 The fundamental symmetry breaking on $\mathscr{B}$ distinguishes the thermodynamic, or power supply, sector Θ from the informative, or computational input/output, sector χ. Black arrow shows the flow of thermodynamic free energy (TFE) from the thermodynamic component Q_Θ to the computational component Q_χ of the internal dynamics H_S. Adapted from [30] Fig. 3, CC-BY license

the remainder of this book. It will be further elaborated to produce computational agents capable of identifying and both preparing and measuring the states of external systems, and hence naturally regarded as "observers" [32].

Box 2.3: The Free Energy Principle

Beginning in the mid-2000s, Karl Friston and colleagues developed a formalism for describing nervous systems [33], organisms [34], and eventually generic, classical dynamical systems [35–37] as approximate Bayesian uncertainty minimizers. The formalism employs a variational free energy (VFE) as an upper bound on Bayesian uncertainty, following an approach originally pioneered by Richard Feynman [38]; it is, therefore, called the *Free Energy Principle* (FEP) formalism and associated modeling framework. The FEP has been widely applied to problems in neuroscience and the broader life sciences, and is increasingly seeing applications in the social sciences. When formulated informally, the FEP simply says that bounded systems act in a way that preserves the integrity of their boundaries.

The FEP applies to all systems that are conditionally statistically independent of their environments; in a classical framework, such systems can be represented as bounded by Markov blankets [39]. The Markov blanket includes states that encode inputs to ("sensations") and outputs from ("actions") the system. These include thermodynamic inputs and outputs [40]; however, these are elided in most FEP applications. A Markov blanket is obviously symmetrical: the "sensations" of a system are the "actions" of its environment and vice-versa. The FEP framework has, therefore, the same basic "picture" of physical interactions as Fig. 2.1. It can, therefore, be given a quantum-theoretic formulation using the language of Hamiltonian operators, holographic screens, and QRFs [41, 42], which we will develop in detail in Chap. 6. In this formulation, the limit of zero VFE on the boundary $\mathscr{B}$ corresponds to the breakdown of separability, i.e. to S and E becoming entangled across $\mathscr{B}$.

Translating a generic quantum-theoretic model into the language of the FEP allows us to view generic quantum systems as approximate Bayesian uncertainty minimizers. The FEP thus provides us with a natural, agentive semantics for generic systems that allows us to view them, in a formally-precise, non-metaphorical, and scale-free way, as the "observer-participants" that John Archibald Wheeler proposed to be both the constituents and the co-creators of "reality" [43].

2.5 QRFs for Time, Memory, and Objects

With the above as background, we can undertake a functional characterization of the QRFs needed to build a system capable of doing experiments and recording their results—the minimal "observer" needed to do science. In the language of the FEP, this is an "active inference agent" capable of both learning about its environment, and provoking its environment to obtain novel and potentially-surprising information [44, 45]. James Hartle refers to such systems as "information gathering and using systems" [46]. These systems are smart enough to learn how to learn, i.e. to learn what parts of their environments evince sufficient regularity, given their QRFs, to be learnable and to focus on these, ignoring (at least for the most part) the unlearnable parts of their environments [47].

2.5.1 Pattern Recognition, Learning, and QRF Commutativity

Let us first consider a simple supervised learning system, e.g. an artificial neural network (ANN) equipped with a back-propagation algorithm [48]. Such systems learn to recognize patterns in their input data by adjusting parameters—connection weights in an ANN—in their pattern-recognition algorithms. In the simplest case, the algorithm learns to recognize one specific input pattern, which we can take to be one specific bit string, and to reject all others. If we consider the inputs to be encoded by $(n - k)$-qubit strings, recognizing one specific input pattern is obtaining an outcome value of $+1$ from a subset of operators $M_k, M_{k+1}, \ldots M_n$, which requires fixing a particular sequence of single-qubit QRFs, i.e. phase angles $\phi_k, \phi_{k+1}, \ldots \phi_n$. The specific pattern $+1_k, +1_{k+1}, \cdots +1_n$ is then obtained by a logical AND operation on the individual outcome values, which can be implemented by an inverse decision tree as shown in Fig. 2.3.

The ordered phases $\phi_k, \phi_{k+1}, \ldots \phi_n$ and the inverse AND tree that encodes, at its vertex C, the desired pattern $+1_k, +1_{k+1}, \cdots +1_n$ together comprise an $(n - k)$-qubit QRF Q_g that specifically recognizes the pattern $+1_k, +1_{k+1}, \cdots +1_n$. This QRF answers a single yes/no question: is the pattern $+1_k, +1_{k+1}, \cdots +1_n$ encoded on some set of $(n - k)$ qubits? A ruler exactly 10 cm long, but with no other markings, similarly answers a single binary question: is this object 10 cm long? A 10 cm ruler can also be used to prepare a length of

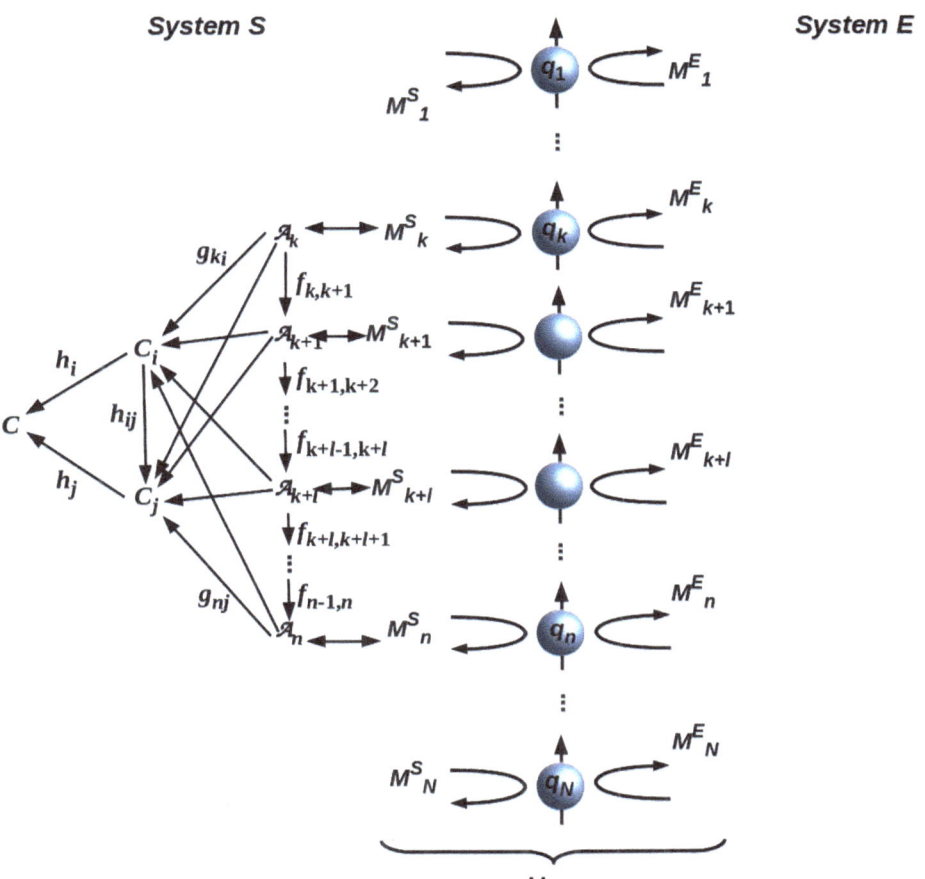

Fig. 2.3 A specific pattern of $(n-k)$ qubit states can be recognized by fixing the phase angles $\phi_k, \phi_{k+1}, \ldots \phi_n$ so that the Hermitian operators $M_k, M_{k+1}, \ldots M_n$ each return an outcome value $+1$, and combining these values with an inverse tree of AND operators. Here the nodes $\mathcal{A}_i$, which will be characterized formally in Chap. 3, recognize outcome values of $+1$ and can be taken to encode the phases $\phi_k, \phi_{k+1}, \ldots \phi_n$, the horizonal arrows implement AND, and the vertical arrows enforce the ordering $k, k+1, \ldots n$. The node $\mathbf{C}$ encodes the desired pattern $+1_k, +1_{k+1}, \cdots + 1_n$; again see Chap. 3 for a formal definition of such operator diagrams in category-theoretic terms. Adapted from [29] Fig. 3, CC-BY license

10 cm, by marking a position for a cut. We will show in Chap. 3, where we provide a fully-general category-theoretic definition of such QRFs, how the action of a multi-qubit QRF such as shown in Fig. 2.3 can be "reversed" to prepare a specific multi-qubit state.

A supervised learner is initialized with some default QRF Q_0, e.g. a random weight matrix in an ANN or a random sequence of phase angles in the notation above. It implements an iterative learning algorithm $\mathcal{L}_i : Q_i \mapsto Q_{i+1}$ such that $Q_i \to Q$ as $i \to m$, where m is the size of the training set. The algorithm $\mathcal{L}_i$ requires an input from the environment for each i, the training input, which we can take to be a binary pattern. If the training input indicates recognition failure at step i, $\mathcal{L}_i$ adjusts the learner's parameters, e.g. the weights or phase angles. We can view the training input as detected by a second QRF Q_{tr}, acting on a distinct sector of $\mathscr{B}$, that produces (or selects) the required parameter adjustment as its observational outcome.

In practice, training inputs alternate with pattern inputs from the training set. It is obvious why this must be the case: Q_i must be well-defined—not in the process of being updated—while operating on the ith input pattern. Hence Q_i and Q_{tr} do not commute: $[Q_i, Q_{tr}] = Q_i Q_{tr} - Q_{tr} Q_i \neq 0$ for all i unless $Q_{tr} = \mathbf{Id}$.

If two operators do not commute, they cannot be implemented by a single quantum process, i.e. by an linear operator $\mathcal{O} : |i\rangle \mapsto |f\rangle$ on a pure state $|i\rangle$. Therefore we have a general result [49, 50]:

Non-commuting QRFs compartmentalize any system that implements them.

Non-commuting QRFs, in other words, factor systems into separable components and additively decompose their internal Hamiltonians. If S implements Q_i and Q_{tr}, for example, we can write $S = S_1 S_2$ where $|S_1 S_2\rangle = |S_1\rangle|S_2\rangle$ and $H_S = H_{S_1} + H_{S_2} + H_{S_1 S_2}$. The interaction $H_{S_1 S_2}$ is defined at a boundary $\mathscr{B}_{S_1 S_2}$ that is *internal* to S. Following the language of [30], we will refer to systems implementing only commuting QRFs as (irreducible) *atomic* systems and systems implementing non-commuting QRFs, i.e. systems with internal boundaries, as *composite* systems.

2.5.2 Time QRFs and Memory

A supervised learning system such as an ANN can be trained to respond specifically to particular patterns in its input, but cannot recognize a specific pattern as something that has been seen and responded to before. This is because it has no memory for specific past events, and no means of measuring time and hence distinguishing a "past" from the present. Such systems cannot, therefore, recognize a pattern as being continuous or even repeated in time.

Recognizing an input as the same as, equivalent to, or even similar to, an input received earlier requires a read/write memory and an ability to measure time. In Bayesian language, the notions of "expectation" or "probability" require memory and time measurement; a system cannot recognize an event as part of an ensemble of events, for example, without the memory and time-labeling resources needed to represent the ensemble. These two resources are inter-defined and inseparable: a remembered event can only be distinguished from a current event if it is marked as being "past" by some timestamp, while a clock—a time QRF—can only be recognized as "ticking" if at least one previous tick has been stored in, and can be retrieved from, a memory. Defining a time QRF therefore both enables and requires the definition of a read/write memory.

Let us begin by defining a *memory* for some system S as a sector Y of S's boundary $\mathscr{B}$ where data can be written and afterwards retrieved, together with a QRF Q_Y that writes to and reads from Y. "Afterwards" here is with respect to the time parameter t_U, the only time parameter currently available. As Y is a sector on $\mathscr{B}$, it is exposed to S's environment E; hence writing on Y is sending an input to E. Memory as defined here is, therefore, fundamentally *stigmergic*: a memory is a mark made on the environment [51]. A memory will be reasonably stable, and hence useful, if the probability that E responds to the memory being written by erasing or altering it is small. The system S clearly cannot measure this probability, but can estimate it by writing multiple copies of the memory and comparing them later, i.e. by employing some kind of error-correcting code (ECC). Making multiple backups of the files on one's computer is an obvious example.

Writing to a memory Y is an action by the QRF Q_Y. In any finite system, actions require finite energy (which must be provided by TFE input from the thermodynamic sector Θ) and finite time (in units of t_U); hence actions can be considered discrete and can be counted. A counter $\mathcal{G}_{ij}$ of actions by Q_Y defines a clock; each transition $i \rightarrow j$ is one "tick" of this clock and defines a time unit t_Y. The operator $\mathcal{G}_{ij}$ is a groupoid operator on the indices i, j [29] (see Appendix A.1 for the definition of a groupoid). If $\mathcal{G}_{ij}^n(i) = i$, the clock is cyclic with period n; the maximum value of n is determined by the coding capacity of Y. Note that $t_Y = \alpha t_U$ for some $\alpha \geq 1$ (typically, $\alpha \gg 1$) with no requirement that α be a constant; t_Y is a time unit for and strictly relative to S. For simplicity, we will assume that S has only one memory and only one clock, and therefore write $t_Y = t_S$ to emphasize the system-relativity of measured time (Fig. 2.4).

Box 2.4: Time, entropy, and the Wick rotation

As discussed in [12], viewing time as not just associated with, but rather defined by the flow of information enables an intuitive understanding of entropy. From S's perspective, E is an information source; N bits of information flow from E to S on each cycle of interaction. Processing these N bits through S's QRFs—using the bits from the Θ sector as fuel—increments S's clock by one unit dt_S. Hence from S's perspective, E radiates information at a rate of N bits per dt_S. As this is all that S observes, S can regard E as an isothermal radiator that gains N bits of entropy per dt_S.

The situation from E's perspective is exactly the same; E sees S radiating information, and hence gaining entropy, at a rate of N bits per dt_E. *With respect to its own time coordinate*, therefore, each system sees the other conforming to the 2nd Law. We can

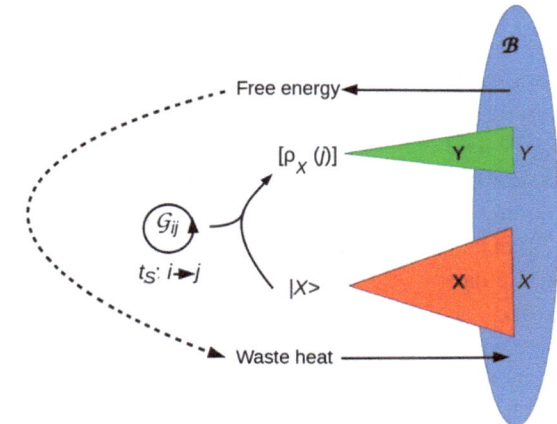

Fig. 2.4 A system S counts recordings of a state measurement $|X\rangle$ to a memory Y using a clock $\mathcal{G}_{ij}$. The record written is represented as a density ρ_X to indicate that it may be coarse-grained. The index value i is available as a timestamp for the ith record written to Y. Adapted from [29] Fig. 9, CC-BY license

depict this situation as in Fig. 2.5, which shows the time arrows t_S and t_E both pointing into their respective systems from $\mathcal{B}$, in line with the flow of detectable information.

We can interpret Fig. 2.5 as depicting an operation: $\mathcal{B}$ reverses t_S to produce t_E. This reversal can be thought of as a rotation by π on an axis in the plane of $\mathcal{B}$. Recall from Chap. 1 that the *Wick rotation* is an equivalence $\iota t/\hbar \leftrightarrow 1/(k_B T)$. As John Baez has discussed in detail [52], the Wick rotation states a formal analogy between the propagator $\exp(-(\iota/\hbar)H)$ and the partition function $\mathrm{Tr}(\exp(1/(k_B T))H)$, where H is the Hamiltonian of a system of interest and Tr is the Trace operator. As such, it relates the conservation of information to the notion of thermal equilibrium or maximal entropy. In Fig. 2.5, we can see $\iota/\hbar$ as a literal rotation acting on t_S as a bit is transferred across $\mathcal{B}$.

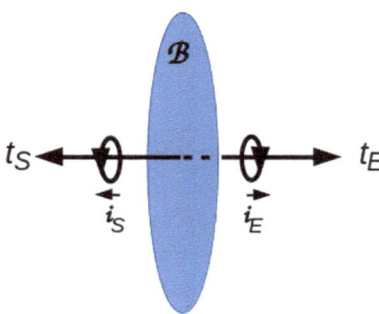

Fig. 2.5 Wick rotation. Adapted from [12] Fig. 6, CC-BY license

2.5.3   QRFs for System Identification and Pointer-State Measurement

Clocks and event memories are the fundamental resources required to identify systems or their states over time and hence to detect and respond to changes in E. We will consider system identification in this section, then discuss space and the detection of motion in the next section.

Let us begin with a thought experiment [53]. Suppose I want Alice to report the observational outcome registered by a particular macroscopic apparatus located in a laboratory filled with lots of apparatus, equipment racks, tables, other experimenters, etc. How much information do I need to give Alice to assure that she reports the outcome from the right apparatus? Clearly, I can give Alice at most a finite description of the apparatus that I want her to report an outcome from. I could instruct her, for example, to locate a black table-top box 30 cm wide and 10 cm high with two analog gauges and a switch, and to report the reading displayed by gauge labeled "B". I could add that the box in question is connected to a counter in the third rack from the right wall. Alice must then enter the laboratory and look for, using the observational means at her disposal, an apparatus matching my finite description.

It is obviously circular to assume that, when Alice enters the laboratory, she can identify the apparatus satisfying the given criteria without having to look at anything else: this is equivalent to assuming that the apparatus is the only thing she is capable of seeing, which would make navigating the cluttered laboratory difficult to say the least. To identify the desired apparatus, Alice must distinguish it, using her criteria and available observables, from everything else in the laboratory, i.e., from all the rest of her environment E. Alice must, in other words, employ her criteria/observables to search E until she finds the apparatus she is looking for. Only then can she measure its state, particularly the value registered by the "B" gauge.

We know from the discussion above how questions like "is this 30 cm wide?" or "does this have two analog gauges?" are answered: each is answered by a specific QRF. An ANN could be trained to recognize black table-top boxes 30 cm wide and 10 cm high with two analog gauges and a switch. An ANN cannot, however, do Alice's job, because an ANN has no event memory. Alice needs to use her available QRFs to find the right box, then read the right gauge, then write both the fact that she found the box and the gauge, and the value indicated on the gauge in memory, then relocate me and read her memory to give me this information.

Let us call the collection of QRFs that Alice uses to find the desired apparatus—all of which must be simultaneously deployable and hence mutually commutative—Q_R or the "reference" QRF for the apparatus, and call the collection of QRFs that Alice uses to determine the state of the gauges Q_P for the "pointer" QRF. What these two QRFs detect is shown in Fig. 2.6. Both QRFs measure components of the state of the apparatus in question: Q_R measured the invariant state components—size, shape, color, etc.—that allow

System *S*

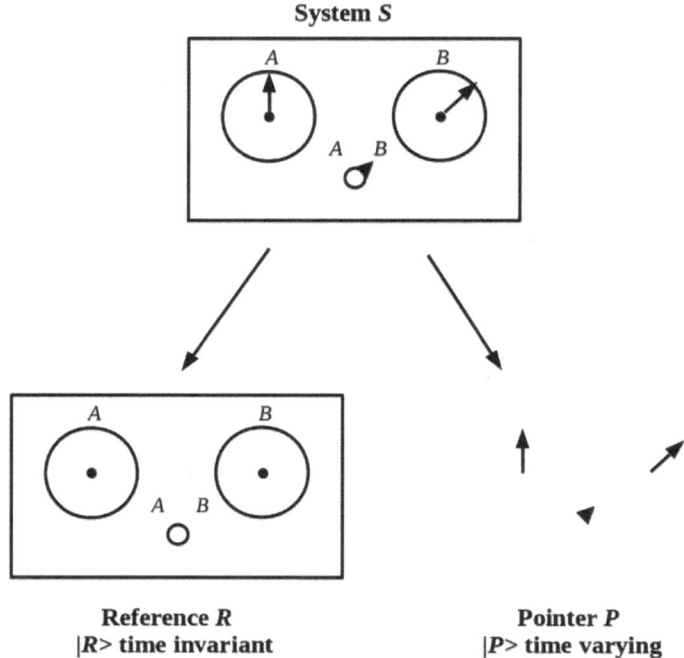

Reference *R*   **Pointer *P***
|*R*> time invariant |*P*> time varying

Fig. 2.6 Identifying reference and pointer states. Adapted from [5] Fig. 2, CC-BY license

identification over time, while Q_P measures the "pointer" components that vary in time and are what is "cf interest" about the apparatus when it is being used.

The QRFs Q_R and Q_P, like any QRFs, read and write information to and from Alice's boundary $\mathscr{B}_A$. Hence the apparatus that they pick out is *defined for Alice* as a pattern (or collection of patterns) of qubit states on $\mathscr{B}_A$ to which Q_R and Q_P selectively respond. Alice treats, and in FEP language represents, the apparatus as a "system" or "object" that is embedded in E, but what Alice detects, and acts upon, is a collection of qubits—a sector— on her boundary. This is, clearly, a simple consequence of the HP: Alice cannot obtain any information from E other than what H_E encodes on $\mathscr{B}_A$. The principle of Decompositional Equivalence stated in Sect. 2.1.2 tells us that what H_E writes on $\mathscr{B}_A$ is independent of how Alice, or any other theorist, might decompose the Hilbert space $\mathcal{H}_E$ into components. Observable "systems" are, therefore, observer- and QRF-dependent by definition. Hence we have:

There are no "objective" observer-independent systems.

Observer-dependence applies to all systems regardless of composition or scale, including laboratory apparatus and other experimenters. "Wigner's friend" scenarios in which another

observer becomes entangled with a system of interest [54, 55] do not pose paradoxes when
the observer-dependence of all systems is taken into account.

Box 2.5: Taking objects for granted

Wojciech Zurek began his influential 1998 review [56] of decoherence theory—the
theory that attempts to explain the "emergence of classicality" within quantum theory
[13]—by pointing out that "it is far from clear how one can define systems given an
overall Hilbert space 'of everything' and the total Hamiltonian" (p. 1794), and ends it
with "a compelling explanation of what the systems are—how to define them given, say,
the overall Hamiltonian in some suitably large Hilbert space—would undoubtedly be
most useful" (p. 1818). Zurek answered his own question by assuming as "axiom(o)"
of quantum theory that "the Universe consists of systems" [57], and proposing the
ideas of the "environment as witness" [58], and "quantum Darwinism" [59] to explain
how observers determine the states of these systems. The underlying intuition is that
observers like us do not interact directly with quantum systems, or even macroscopic
apparatus, but rather interact with the ambient environment, e.g. the ambient photon
field. Alice observes S, for example, by interacting with E_{amb}, which "witnesses" and
encodes information about the state $|S\rangle$ or, if coarse-graining is included in the picture,
the density ρ_S of S. The state information that E_{amb} can encode is limited by the
interaction $H_{SE_{amb}}$; E_{amb} can only encode eigenstates of this interaction. Hence Alice
can only observe eigenstates of $H_{SE_{amb}}$. States of S "compete" to be eigenstates of
$H_{SE_{amb}}$, and only those actually "selected" by $H_{SE_{amb}}$ are observable by Alice.

This picture raises an obvious question: Alice's environment E_A presumably contains lots of systems besides S. When Alice goes to look for the apparatus with the
gauge to be read, it is in a cluttered laboratory. How does Alice know, when interacting with E_{amb}, to look for the information that E_{amb} encodes about S? And even if
she knows to look for it, how does she distinguish the information that E_{amb} encodes
about S from the information that E_{amb} encodes about X or Y or Z? There are only
two possible answers to these questions. Either E_{amb} only encodes information about
S, or Alice already knows what information to look for and how to find it, i.e. has
a reference QRF Q_R for S. The first answer makes decompositions of E_A that pick
out S metaphysically "special" and hence violates Decompositional Equivalence. The
second answer renders the whole story circular [60].

Philosophers have long sought a first-principles, observer-independent (or "objective") ontology, a theory that explained the "furniture of the world" [61] without talking
about us. Many physicists have taken on this quest, or at least professed faith in its
eventually successful outcome. Albert Einstein's famous critique of quantum theory
[62] rested on his belief that there must be "elements of physical reality" independent
of our or any other systems' observations. John Stewart Bell fiercely resisted the idea
that "observers" or "measurement" should play any role in physical theory [63]. The
popular idea that particular macroscopic structures—planets, stars, galaxies, organ-

isms, etc.—somehow "emerge" from the physics of the Big Bang or the Inflationary Epoch [64, 65] reflect this idea that ontology is "given" by physics, and hence can be taken for granted by observers.

The picture we have painted in this chapter obviously contradicts this idea of an observer-independent ontology. Indeed it illustrates a physics with no ontology at all, a physics in which boundaries—including the boundaries around observers—are merely stipulated and have no effect at all on "the overall Hamiltonian" H_U. Every boundary bounds an observer, and determines what information that observer can obtain, but the fact that an observer has obtained some information is only important if that observer's behavior, and the behavior of that observer's environment, are what is of interest. Hence the physics illustrated here is a physics of *information exchange* and *computation*. It is about "things" only to the extent that "things" are what observers model themselves, and each other, as talking about.

This picture challenges us to ask, *within physics*, how observers identify the systems they identify, how they implement memory and computation, and how they manage to communicate. Such questions are often dismissed as outside of physics altogether [57, 66]. We see them, instead, as questions at the very heart of physics.

2.6   Communicating Observers and Spacetime

Suppose Alice and Bob can exchange classical messages, e.g. can talk to each other or send each other emails. In the notation used above, this means that Alice and Bob are separable systems (all joint states $|AB\rangle$ factor as $|A\rangle|B\rangle$) that both interact with some environment E, and that E implements causal processes f and g—processes that require some finite time according to E's internal clock—such that (some components of) $|A\rangle = f(|B\rangle)$ and (some components of) $|B\rangle = g(|A\rangle)$. Let us also suppose that Alice and Bob have something to communicate about, i.e. suppose that Alice and Bob have boundary sectors X_A and X_B, the states of which influence, and are influenced by, some component X of E. Let us suppose, finally, that their observations and manipulations of their respective boundary sectors X_A and X_B are all that they communicate about. In this case, we can represent their interactions as in Diagram (2.4). Here the pair of functions (f, g) implements a classical channel between Alice and Bob, the component X of E implements a quantum channel between Alice and Bob, and Alice and Bob can be described as implementing QRFs Q_A and Q_B, respectively, that write observations of X_A and X_B, respectively, to the classical channel as messages, and write contents of messages to X_A and X_B, respectively, as state preparations. Alice and Bob communicate in this way when, for example, they share a laboratory apparatus—X—that they both observe, manipulate, and talk about.

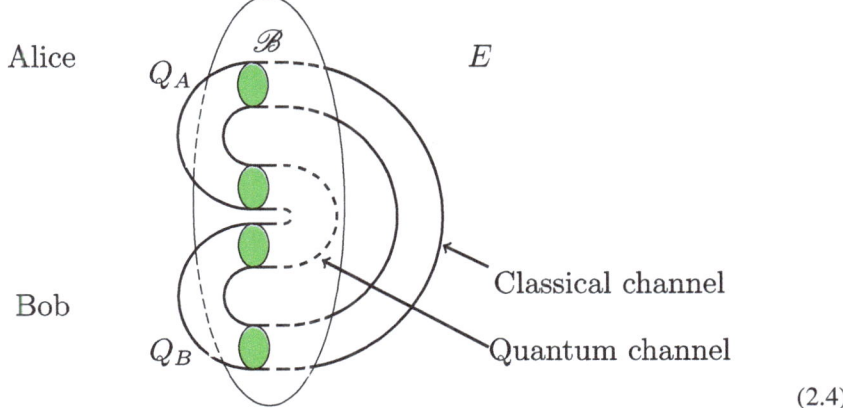

$$(2.4)$$

Interactions between communicating observers that have the form shown in Diagram (2.4) are *local operations, classical communication* (LOCC) protocols [67], which will be discussed in more detail in Chap. 7. A canonical Bell/EPR experiment [25] is a LOCC protocol [68]: Alice and Bob, independently and locally in their own laboratories, manipulate and then observe a (finite sequence of) entangled quantum states, then pool their outcome statistics to look for violations of the CHSH inequality. The entangled states are quantum channels; the pooled statistical analysis requires classical communication. Classical communication is also required before the experiment, when Alice and Bob agree to both measure spin, to employ as far as possible the same z axis (e.g. the Earth's gravitational field), and to synchronize their clocks. Each of these pre-experiment coordination exercises is also a LOCC protocol; synchronizing their clocks requires interacting with a time standard located in E, i.e. a quantum channel.

Multi-observer quantum Darwinism is also a LOCC protocol [68]: Alice and Bob can both extract the same state information about some system S embedded in E by interacting locally with E, provided they agree in advance to deploy effectively-equivalent reference (for system identification) and pointer (for state measurement) QRFs. Note that Alice and Bob can transfer quantum information via S only if their QRFs do not commute, i.e. only if Alice's action on S affects Bob's outcomes and vice-versa. Hence ideal quantum Darwinism— Alice and Bob seeing exactly the same outcomes—requires that $Q_A = Q_B$, rendering Alice and Bob entangled. In practice, therefore, quantum Darwinism requires coarse-graining, i.e. it enables Alice and Bob to agree about ρ_S, at some finite resolution, while deploying effectively-equivalent but non-identical QRFs.

The quantum channel in any LOCC protocol functions as a *quantum error correcting code* (QECC) [68], again as will be discussed in detail in Chap. 7. In the language of [27], the perturbations B_α the channel must protect against are the interactions $H_{C\bar{C}}$ between the channel degrees of freedom C and another degrees of freedom $\bar{C}$ of E. Any such interactions introduce decoherence and hence quantum information loss in C. Adding redundancy to C

is adding dimensionality to the codespace of the QECC that C implements; the redundant qubits effectively serve to "insulate" C against the action of $H_{C\bar{C}}$, and hence to prevent decoherence.

Box 2.6: Spacetime is emergent

Wheeler declared in [43] that neither time nor space could be fundamental in any fully-consistent quantum-theoretic model of the universe. We have seen above how observable time can be constructed from information flow. With the discovery of the duality between a gravitational theory of an Anti-deSitter bulk and a conformal field theory on its boundary (AdS/CFT duality [69]), a rapidly-expanding literature has explored multiple ways of constructing spacetimes compliant with Einstein's equations from quantum interactions, particularly entanglement relations; see [70, 71] for relatively informal reviews and [72] for an analysis of various approaches in the language used here.

Intuitively, spacetime provides a "place to put" redundant instances of some identified system. If the same system S can be identified by reference QRFs that detect different reference states, then a redundancy resource that allows representations of S must represent S as invariant across maps between alternative S-identifying reference states. Human object identification QRFs are (ideally) equivalent under transformations of identified systems by operators in the Poincaré Group: spatial translations, rotations, and boosts. The idea that "objects" or "particles" are representations of the Poincaré Group was originally proposed by Eugene Wigner [73]. Spacetime is exactly the resource that supports this symmetry.

That space, as well as time, should be an observer-relative construct is not surprising. If identified systems are sectors of $\mathscr{B}$—the domains on $\mathscr{B}$ of system-identification QRFs—then "space" is a QRF that assigns a geometry to $\mathscr{B}$. In mammals, this QRF is implemented by the hippocampus [74], and must at least in part be learned through experience, e.g. by motor babbling. There is no reason to believe that all organisms represent "space" in the same way, or even that all organisms represent space at all.

2.7   Non-Go Theorems for Generic Quantum Systems

As discussed in Chap. 1, Edward Moore proved in 1956 that no finite sequence of finite input/output experiments on a black box could determine the "machine table" or inner workings of the black box [75]. A black box is just a physical system that can be interacted with only via inputs and outputs encoded on its boundary. As Fig. 2.1 makes obvious, the HP renders all bounded physical systems black boxes. Moore's Theorem states, therefore, what we already know from Sect. 2.2.2: *S can only obtain from E the information E encodes on its boundary, and vice versa.*

This simple observation yields a number of powerful and, in some cases, counter-intuitive "no-go" results, all of which can be viewed as quantum-informational elaborations of Moore's Theorem [76].

Theorem 2.7.1 ([76] Thm. 1) *Let S be a finite system and Q be a QRF implemented by* H_S. *The following statements hold:*

1. *S cannot determine, by means of Q, either Q's dimension* $\dim(Q)$, *Q's associated sector dimension* $\dim(\mathrm{dom}(Q))$, *or Q's complete I/O function.*
2. *S cannot determine, by means of Q, the dimension, associated sector dimension, or I/O function of any other QRF* Q' *implemented by S.*
3. *S cannot determine, by means of Q, the I/O function or dimension of any QRF* Q' *implemented by any other system* S', *regardless of the relation of S to* S', *from* $S' = S$ *to* $S' = E$, *inclusive.*
4. *Let* $S = S_i S_j$, *in which case* $E_i = E S_j$. *Then* S_i *cannot determine, by means of a QRF* Q_i, *the I/O function or dimension of any QRF* Q_j *implemented by* S_j.

Proof See [76]. All clauses follow from the inability to specify H_S or H_E given only H_{SE}, or in particular, the finite set of bits encoded on some observable sector of $\mathscr{B}$. □

Perhaps even more important for our purposes is another straightforward consequence of the definition of $\mathscr{B}$:

Theorem 2.7.2 ([77] Cor. 3.1) *No finite system S can measure the entanglement entropy* $\mathcal{S}(|SE\rangle)$ *across the boundary* $\mathscr{B}$ *that separates it from its environment E.*

Proof Trivially, from the inability of S to access either its own state $|S\rangle$ of its environment's state $|E\rangle$. □

No system, therefore, can infer by observation that it is disentangled from its environment. Consequently no system can infer by observation that it has a free choice of QRFs.

References

1. Nielsen, M. A., Chuang, I. L. (2000) Quantum Information and Quantum Computation. Cambridge University Press, New York, NY, USA.
2. Zanardi, P. Virtual quantum subsystems. Phys. Rev. Lett. 2001, 87, 077901; Zanardi, P.; Lidar, D.A.; Lloyd, S. Quantum tensor product structures are observable-induced. Phys. Rev. Lett. 2004, 92, 060402.
3. Bell, J.S. On the Einstein–Podolsky-Rosen paradox. Physics 1964, 1, 195-200.

4. Clauser, J.F.; Horne, M.A.; Shimony, A.; Holt, R.A. Proposed experiment to test local hidden-variable theories. Phys. Rev. Lett. 1969, 23: 880-884.
5. Fields, C.; Glazebrook, J.F. Representing measurement as a thermodynamic symmetry breaking. Symmetry 2020, 12, 810.
6. Tribus, M.; McIrving, E. C. (1971). Energy and Information, Scientific American, 225: 179-188.
7. Bekenstein, J. D. Black holes and entropy. Physical Review D 1973; 7(8):2333-2346.
8. 't Hooft, G. Dimensional reduction in quantum gravity. in: A. Ali, J. Ellis, S. Randjbar-Daemi (Eds.), Salamfestschrift: A Collection of Talks from the Conference on Highlights of Particle and Condensed Matter Physics ICTP, Trieste, Italy, 8-12 March 1993. World Scientific, Singapore, 1994. pp. 284-296.
9. Susskind, L. The world as a hologram. Journal of Mathematical Physics 1995; 36(11):6377-6396.
10. Bousso, R. The holographic principle. Reviews of Modern Physics 2002; 74(3):825-874.
11. Addazi, A.; Chen, P.; Fabrocini, F.; Fields, C.; Greco, E.; Lutti, M.; Marcianò, A.; Pasechnik, R. Generalized holographic principle, gauge invariance and the emergence of gravity à la Wilczek. Front. Astron. Space Sci. 2021; 8: 563450.
12. Fields, C. Glazebrook, J. F.; Marcianò, A. The physical meaning of the Holographic Principle. Quanta 2022; 11:72-96.
13. Schlosshauer, M. Decoherence and the Quantum to Classical Transition; Springer: Berlin, Heidelberg, Germany, 2007.
14. Pegg, D.; Barnett, S.; Jeffers, J. Quantum theory of preparation and measurement. Journal of Modern Optics 2002; 49(5-6):913-924.
15. Landsman, N.P. Between classical and quantum. In Handbook of the Philosophy of Science: Philosophy of Physics; Butterfield, J., Earman, J., Eds.; Elsevier: Amsterdam, The Netherlands, 2007; pp. 417-553.
16. Cabello, A. Interpretations of quantum theory: A map of madness. In: Lombardi, O., Fortin, S., Holik, S., López, C. (Eds.) *What is Quantum Information?* Cambridge University Press, Cambridge, UK, 2017, pp. 138-140.
17. Fuchs, C. Quantum mechanics as quantum information (and only a little more). Preprint arxiv:quant-ph/0205039v1, 2002.
18. Mermin, N. D. There is no quantum measurement problem. Physics Today 2022; 75(6): 62-63.
19. Mermin, D. Making better sense of quantum mechanics. Reports on Progress in Physics 2017; 82(1), 012002.
20. Bartlett, S.D ; Rudolph, T.; Spekkens, R.W. Reference frames, superselection rules, and quantum information. Rev. Mod. Phys. 2007, 79: 555-609.
21. Wootters, W. T.; Zurek, W. H. A single quantum cannot be cloned, Nature 1982; 299: 802-803.
22. Fields, C.; Marcianò, A. Sharing nonfungible information requires shared nonfungible information. Quantum Reports 2019; 1: 252-259.
23. Conway, J.; Kochen, S. The free will theorem. Foundations of Physics 2006; 36:1441-1473.
24. Conway, J.; Kochen, S. The strong free will theorem. Notices of the AMS 2009; 56:226-232.
25. Aspect, A.; Graingier, P.; Roger, G. Experimental realization of Einstein-Podolsky-Rosen-Bohm *Gedankenexperiment*: A new violation of Bell's inequalities. Physical Review Letters 1982; 49(2): 91-94.
26. Jacques, V.; Wu, E.; Grosshans, F.; Treussart, F.; Grangier, P.; Aspect, A.; Roch, J.-F. Experimental realization of Wheeler's delayed-choice Gedankenexperiment. Science 2007; 315: 966-968.
27. Knill, E.; Laflamme, R. Theory of quantum error-correcting codes, Phys. Rev. A 1977; 55: 900-911.
28. Parrondo, J.M.R.; Horowitz, J.M.; Sagawa, T. Thermodynamics of information. Nat. Phys. 2015, 11, 131-139.
29. Fields, C.; Glazebrook, J.F.; Marcianò, A. Reference frame induced symmetry breaking on holographic screens. Symmetry 2021, 13, 408.

30. Fields, C.; Goldstein, A.; Sandved-Smith, L. Making the thermodynamic cost of active inference explicit. Entropy 2024; 26, 622.

31. Fields, C., Glazebrook, J. F. and Marcianò, A. Sequential measurements, topological quantum field theories, and topological quantum neural networks. Fortschritte der Physik 2022; 70: 202200104.

32. Fields, C. If physics is an information science, what is an observer? Information 2012; 3: 92-123

33. Friston, K. The free-energy principle: A unified brain theory? Nat. Rev. Neurosci. 2010, 11, 127-138.

34. Friston, K. Life as we know it. J. R. Soc. Interface 2013, 10, 20130475.

35. Friston, K.J. A free energy principle for a particular physics. arXiv 2019, arXiv:1906.10184.

36. Ramstead, M.J.; Sakthivadivel, D.A.R.; Heins, C.; Koudahl, M.; Millidge, B.; Da Costa, L.; Klein, B.; Friston, K.J. On Bayesian mechanics: A physics of and by beliefs. Interface Focus 2022, 13, 20220029.

37. Friston, K.J.; Da Costa, L.; Sakthivadivel, D.A.R.; Heins, C.; Pavliotis, G.A.; Ramstead, M.J.; Parr, T. Path integrals, particular kinds, and strange things. Phys. Life Rev. 2023, 47, 35-62.

38. Feynman, R.P. Statistical Mechanics; Benjamin: Reading, MA, USA, 1972.

39. Pearl, J. Probabilistic Reasoning in Intelligent Systems: Networks of Plausible Inference; Morgan Kaufmann: San Mateo, CA, USA, 1988.

40. Sengupta, S.; Stemmler, M.B.; Friston, K.J. Information and efficiency in the nervous system–A synthesis. PLoS Comp. Biol. 2013, 9, e1003157.

41. Fields, C.; Friston, K.J.; Glazebrook J.F.; Levin M. A free energy principle for generic quantum systems. Prog. Biophys. Mol. Biol. 2022, 173, 36–59.

42. Fields, C.; Fabrocini, Friston, K.; Glazebrook, J.F.; Hazan, H.; Levin, L.; Marcianò, A. Control flow in active inference systems, Part I: Formulations of classical and quantum active inference. IEEE Trans. Mol. Biol. Multi-Scale Commun. 2023, 9, 235–245.

43. Wheeler, J.A. Information, physics, quantum: The search for links. In Complexity, Entropy, and the Physics of Information; Zurek, W.H., Ed.; Westview: Boulder, CO, USA, 1990; pp. 3–28.

44. Friston, K.; FitzGerald, T.; Rigoli, F.; Schwartenbeck, P.; Pezzulo, G. Active inference: A process theory. Neural Comput. 2017, 29, 1-49.

45. Ramstead, M.J.D.; Constant, A.; Badcock, P.B.; Friston, K.J. Variational ecology and the physics of sentient systems. Phys. Life Rev. 2019, 31, 188-205.

46. Hartle, J.B. The quasiclassical realms of this quantum universe. Found. Phys. 2011, 41, 982-1006.

47. Gottlieb, J.; Oudeyer, P.-Y.; Lopes, M.; Baranes, A. Information-seeking, curiosity, and attention: Computational and neural mechanisms. trends in Cognitive Sciences 2013; 17(11): 585-593.

48. Rumelhart, D.E.; Hinton, G.E.; Williams, R.J. Learning representations by back-propagating errors. Nature 1986, 323, 533-536.

49. Fields, C.; Glazebrook, J. F. Information flow in context-dependent hierarchical Bayesian inference. Journal of Experimental and Theoretical Artificial Intelligence 2022; 34: 111-142

50. Fields, C. The free energy principle induces compartmentalization. Biochemical and Biophysical Research Communications 2024; 723: 150070

51. Heylighen, F. Stigmergy as a universal coordination mechanism I: Definition and components. Cognitive Systems Research 2016; 38: 4-13.

52. Baez, J. Getting to the bottom of Noether's Theorem. arxiv:2006.14741, 2000.

53. Fields, C. Some consequences of the thermodynamic cost of system identification. Entropy 2018; 20: 797.

54. Wigner, E. P. Remarks on the mind–body question. in Symmetries and Reflections, pp. 171–184. (Indiana University Press, 1967).

55. Frauchiger, D.; Renner, R. Quantum theory cannot consistently describe the use of itself. Nature Communications 2018; 9: 3711.

56. Zurek, W. H. Decoherence, einselection and the existential interpretation (the rough guide). Phil. Trans. R. Soc. Lond. A (1998) 356: 1793-1821.
57. Zurek WH Decoherence, einselection, and the quantum origins of the classical. Rev Mod Phys 2003; 75:715-775
58. Ollivier H, Poulin D, Zurek WH Environment as a witness: selective proliferation of information and emergence of objectivity in a quantum universe. Phys Rev A 2005; 72:042113,
59. Zurek WH Quantum Darwinism. Nat Phys 2009; 5:181-188.
60. Fields, C. On the Ollivier-Poulin-Zurek definition of objectivity. Axiomathes 2014, 24: 137-156.
61. Bunge, M. The Furniture of the World. Reidel, 1979.
62. Einstein, A.; Podolsky, B.; Rosen, N. Can quantum mechanical description of physical reality be considered complete? Physical Review 1935; 47: 777-780.
63. Bell, J. S. Against 'measurement'. Physics World 1990; 3(8): 33-41.
64. Weinberg, S. The First Three Minutes. Basic Books, 1977.
65. Hawking, S. A Brief History of Time. Bantam Dell, 1988.
66. Fuchs, C. QBism, the perimeter of Quantum Bayesianism. arxiv:1003.5209v1, 2010.
67. Chitambar E. Leung, D., Mančinska, L., Ozols. M. and Winter, A. Everything you always wanted to know about LOCC (but were afraid to ask). Commun. Math. Phys. 2014; 328: 303-326.
68. Fields, C., Glazebrook, J. F. and Marcianò, A. Communication protocols and QECCs from the perspective of TQFT, Part I: Constructing LOCC protocols and QECCs from TQFTs. Fortschritte der Physik 2024; 72: 202400049
69. Maldacena, J. The large N limit of superconformal field theories and supergravity. Advances in Theoretical and Mathematical Physics. 1998; 2(4): 231-252.
70. Bain, J. Spacetime as a quantum error correcting code? Stud. Hist. Phil. Sci. B 2020; 71: 26-36.
71. van Raamsdonk, M. Spacetime from bits. Science 2020; 370: 198-202.
72. Fields, C., Glazebrook, J. F. and Marcianò, A. Communication protocols and QECCs from the perspective of TQFT, Part II: QECCs as spacetimes. Fortschritte der Physik 2024; 72: 202400050.
73. Wigner, E. P. On unitary representations of the inhomogeneous Lorentz group. Annals of Mathematics 1939; 4091): 149-204.
74. Danjo, T. Allocentric representations of space in the hippocampus. Neuroscience Research 2020; 153: 1-7.
75. Moore, E.F. Gedankenexperiments on sequential machines. In Autonoma Studies; Shannon, C.W., McCarthy, J., Eds.; Princeton University Press: Princeton, NJ, USA, 1956; pp. 129-155.
76. Fields, C.; Glazebrook, J. F.; Levin, M. Principled limitations on self-representation for generic physical systems. *Entropy* 2024, *26*, 194.
77. Fields, C.; Glazebrook, J. F. Separability, contextuality, and the quantum Frame Problem. International Journal of Theoretical Physics 2023; 62: 159.

Category-Theoretic Approach to Distributed Information Flow

3

This chapter develops a representation of distributed information flow couched in the language of category theory. This representation provides a natural formalism for specifying the functions computed by QRFs, and therefore for showing how the dynamics of generic physical systems implement information-processing operations, i.e. computations.

3.1 Basic Aspects of Information Theory

The history and applications of the theory of information to computational and communicating systems is both extensive and varied. Here, we will recall some of the basic features of the theory from quantitative, qualitative-semantic, and physical (thermodynamic) perspectives. We will commence with the latter.

3.1.1 Information and Thermodynamics

As we surveyed in Chap. 1, and briefly recall here, the link between information, or entropy, and thermodynamics was originally pioneered by Boltzmann [1], who was the first to observe that reduction in uncertainty entails energetic costs. Boltzmann's ideas can be tied in with the classical "Maxwell's Demon" thought experiment in relationship to the Second Law of Thermodynamics, following which a number of devices, such as the Szilard Engine, were introduced to both confirm and advance Boltzmann's original findings. It was thanks to C. E. Shannon's ground-breaking work that close connections between uncertainty reduction and communication, i.e. information transfer, were established [2, 3]. Shannon's theory, the development of cybernetics as an abstract, information-theoretic approach to engineering [4],

© The Author(s), under exclusive license to Springer Nature Switzerland AG 2026
C. Fields and J. Glazebrook, *Distributed Information and Computation in Generic Quantum Systems*, Synthesis Lectures on Engineering, Science, and Technology,
https://doi.org/10.1007/978-3-031-97263-8_3

the realization that computers were not just number-crunchers but general-purpose inference devices [5], McCulloch and Pitts' model of neurons as information processors [6], and the use of informational language in biology [7] all contributed to a gradual re-interpretation of physics as a theory of information, as originally suggested by R. Feynman [8], and later elaborated by J. Wheeler [9]. An early development was *Landauer's Principle* [10] relating memory-erasure to energy costs, and C. H. Bennett's extensive work on the thermodynamics of computation [11, 12], the main thrust of the conclusions pointing to information and free energy as formally equivalent. Reviews surveying the landscape of the subject through its history and methods may be found in [3, 8, 9, 11–13].

We proceed by briefly recalling the very basics established in Shannon's theory, namely *Entropy* (or *Uncertainty*) and *Mutual Information*. The entropy $S(X)$ of a random variable X in terms of its probability distribution $p_X(x)$, is given by

$$S(X) = S[p_X] - \sum_x p_X(x) \log p_X(x) \tag{3.1}$$

which, since Shannon, has been viewed as a quantitative measure of information. If the state X of a system involves measurement of a quantity Y, which when measured yields information about X, that in turn decreases the degree of uncertainty. *Mutual information*, denoted $I(X; Y)$, is given by

$$I(X; Y) \equiv S(X) - S(X|Y) \tag{3.2}$$

This specifies the amount of information that Y provides about X, where $S(X|Y)$ is the uncertainty of the posterior probability distribution $p_{X|Y}(x|y)$ averaged over possible outcomes $p_Y(y)$, as given by

$$S(X|Y) \equiv \sum_y p_Y(y) \Big[- \sum_x p_{X|Y}(x|y) \log p_{X|Y}(x|y) \Big]$$
$$= - \sum_{x,y} p_{XY}(x, y) \log p_{X|Y}(x|y) \tag{3.3}$$

In terms of the Bayes' formula for relating joint and conditional probabilities, $p_{XY}(x, y) = p_{X|Y}(x|y)p_Y(y)$, (3.2) can be expressed as:

$$I(X; Y) = \sum_{x,y} p_{XY}(x, y) \log \frac{p_{XY}(x, y)}{p_X(x)p_Y(y)} \tag{3.4}$$

(for further details see [2, 3, 14] which include outlines of Shannon's acclaimed results on Channel Capacity and Rate Distortion).

Box 3.1: What is probability?

Are probabilities real features of the world, or do they just describe experience? Is there such a thing as chance—or indeterminism, or randomness—or are there just such things as ignorance and observational limits? Philosophers, mathematicians, and physicists have debated these questions for centuries, with no agreement in sight; see [15–17] for recent surveys.

Given our emphasis on the observer-relativity of observational outcomes, objects, and even spacetime in Chap. 2, it will not be surprising that we treat probabilities as observer-relative. Indeed by defining "information" in terms of probability, and then equating information and entropy, we are required to treat probabilities as observer relative since we have defined entropy as observer-relative. David Mermin [18] provides a spirited defense of this position.

How one interprets probabilities makes no difference to their formal description. Viewing them as observer-relative does, however, forbid a certain kind of laziness, requiring the questions "by whom?" and "using what QRFs?" and "in what context?" whenever a probability is said to be "measured".

In 1961, Landauer proposed a limiting principle asserting that the minimum possible amount of energy required to erase one bit of information, is

$$E = k_B T \ln 2 \tag{3.5}$$

where we recall Boltzmann's constant $k_B = 1.38 \times 10^{-23}$ J/K, and T is the absolute temperature [10] (reviewed in [12, 19]). Alternatively, if ΔE_{env} denotes energy dissipated into the environment, and ΔS_{sys} is the thermodynamic entropy equivalent to information erased from the system's memory, then in the presence of some heat reservoir:

$$\Delta E_{\text{env}} \geq k_B T \Delta S_{\text{sys}} \tag{3.6}$$

If the system is, furthermore, noise free, and efficiently records information to the extent that only $k_B T \ln 2$J is dissipated per bit overwritten, then we have

$$N \leq \frac{\Delta E_{\text{env}}}{k_B \ln 2} \tag{3.7}$$

where N is a dimensionless number depending on memory erasure (reviewed in [19, 20]). Equation (3.5) involving entropy implies an energy cost. This energy cost, as independent of temperature, can be avoided if the reservoir is designed in a way that allows paying an equivalent cost in units of angular momentum or some other conserved quantity [21].

Likewise, following an original example introduced by Feynman [8], there is a direct relationship of energies and informational states to free energy. Given an informational state ρ with Hamiltonian H depending on an index m, with partition function Z_M, *non-equilibrium free energy* is given by [14, Sect. 5]:

$$\mathfrak{F}(\rho, H) = \sum_m p_m F_m - T S(p_m) \tag{3.8}$$

where $F_m = -k_B T \ln Z_m$ is restricted free energy, and $S(p_m)$ is the Shannon entropy of the probabilistic informational state p_m (as dependent on history, measurements, etc.). If the system is in global equilibrium, then it can be seen that (3.8) becomes $\mathfrak{F}(\rho, H) = -kT \ln Z$, where $Z = Z_m$ is the global partition function. For a memory that is symmetric, whereby $F_m = F_{\text{local}}$ is the same across informational states, then:

$$\mathfrak{F}(\rho, H) = F_{\text{local}} - T S(p_m) \tag{3.9}$$

If a symmetric memory is manipulated such that the system is transformed isothermally from distributions $p_m \mapsto p'_m$, and the initial and final restricted free energies are the same, then the work W needed to complete the process complies with the following bound:

$$W \geq T[S(p_m) - S(p'_m)] \tag{3.10}$$

Note that (3.10) shows that work is necessary to order to maintain a memory as $S'(p_m) < S(p_m)$, whereas work is extractable when $S'(p_m) > S(p_m)$. Equation (3.10) generalizes Landauer's erasure principle in (3.5), for which the work necessary is $W \geq k_B T \ln 2$.

3.1.2 Shannon, Dretske, and Distributed Systems

The informal concept of information has a long history of interpretations and accompanying philosophical controversy (see [22] for a summary). Shannon's information theory, as briefly summarized in Sect. 3.1.1, provides a *quantitative, semantics-free* answer to the question of *how much* information can be transferred via a given communication protocol or device [2, 3]. Shannon's theory intentionally ignores the question of what the transmitted information is *about*. Hence while the theory employs the notions of probability and uncertainty, it says nothing about how an agent can employ information received through a channel to update a probability or reduce an uncertainty.

Fred Dretske [23] addressed these questions by proposing a *qualitative* information theory that focuses on the *semantic content* of the transmitted information. Unlike Shannon's theory, Dretske's information theory is concerned with how information is used to construct *knowledge* [24]. We can account for the basis of Dretske's approach in several informal steps, following [25, Ch. 1] as a review of the essential ingredients of [23]. Dretske begins with intuitions along the lines of: Alice knows that p if she believes that p, and her believing that p carries the information that p. In a similar vein: "r's being F carries the information that s is G". So we may inquire about the information that is conveyed by s being G with some grammatical regularity. An explanation can be summarized using conditional probabilities by Dretske's notion of Information Content:

To a person with prior knowledge k, r being F carries the information that s is G, if and only if the conditional probability of s being G given that r is F, is 1 (and less than 1 given k alone).

This is followed through by Dretske's 'Xerox Principle', which states that the operation "carries the information" is transitive:

If r being F carries the information that s is G, and s being G carries the information that t is H, then r being F carries the information that t is H

(see [25, pp. 15–16]). Dretske's notion of "carrying information" is explicitly causal: the means by which probabilities account for information is, for Dretske, analogous to how the former can account for causality. Care must be taken, however, in interpreting information flow causally: the past can carry information about the future, in the sense that the past can predict some aspects of the future, but the future does not, for Dretske, cause the past.

Box 3.2: Semantics and pragmatics

Shannon's information theory was developed in a communications-engineering context, as a way of quantifying the bandwidth of potentially noisy channels. Dretske's information theory, in contrast, builds on centuries of philosophical theories of knowledge—i.e. of epistemology—and on theories of how symbols acquire meanings that began to be developed in the late 19th century. They have, therefore, very different motivations and foci: Shannon's theory is about fidelity, while Dretske's theory is about truth.

Intuitively, a sentence is true if what it states is in fact the case; "snow is white" is true because snow is, as a matter of fact, white. Away from such "obvious" cases, truth has traditionally been taken to follow from sound deduction; i.e. true conclusions follow from true premises and valid logic. This intuitive picture was shattered in 1931 by Kurt Gödel's proof [26] that no formal system that could express elementary arithmetic could be both consistent and complete. Even elementary arithmetic, in other words, includes true statements that are not provable, false statements that are provable, or both. The foundational connection between deduction and truth had been cut.

The resulting scramble to resurrect a workable theory of truth—and hence a workable theory of the meanings of symbols and how combining these meanings could yield truth—consumed much of the second half of the 20th century. It yielded two main approaches: *representational* theories in which symbols are connected to objects, processes, properties, or facts by some kind of formalizable mapping, and *pragmatic* theories in which the abstract idea of "meaning" is more or less abandoned in favor of a focus on how words are used in communicative contexts. Wittgenstein, for example, championed a focus on the actual usage of language in his *Philosophical Investigations* [27]. This pragmatic strategy is epitomized by Large Language Models (LLMs) such as the GPT series [28]; LLMs are statistical models of language usage that, at least in the eyes of some theorists, have no understanding of language at all [29].

Dretske's information theory is explicitly representational, with causation as the mapping between words and their referents. The choice of causation as the fundamental relationship underlying semantics, and therefore truth, is motivated in part by the arguments of Saul Kripke [30] and others, that the relation between some terms—particularly proper names—and the objects to which they refer is a matter of objective fact, independent of what any user of such terms knows or believes. Carried over to the notion of truth, this means that snow being white as a physical fact *causes* "snow is white" to be objectively true, independently of questions of usage.

Representational semantics need not assume observer-independence; multiple observers can offer multiple, distinct interpretations of any physical system, and hence of any array of "objects" and "processes" in their observed worlds, as the discussion of computational interpretations in Chap. 1 makes clear. Gregory Bateson's proposal that the "differences that make a difference" for a particular observer define what is meaningful for that observer [31] localizes meaning to each particular observer's *umwelt*. As emphasized by Maturana and Varela [32], and many others, what "makes a difference" to an observer is *actionability*, either now, or in the future. Hence observer-relative representational, or *interpretative*, semantics are also observer-relative pragmatic semantics: what is important about language, or about any element of language, is the use to which it can be put [33].

3.2   The Distributed Information Flow Theory of Barwise-Seligman

Barwise and Seligman [25] extended the theories of both Shannon and Dretske into a theory of logically regulated information flow that recognizes both the role of the environment as a causal provider of information, and the role of the observer as an interpreter of information. For example, quoting from [25, p.7]:

> .. when one looks at nonlinguistic forms of communication, and the many other phenomena that we have listed as examples of information flow, the spatial and temporal relationships between parts of a system cannot be ignored. The very term "information flow" presupposes, albeit metaphorically, a spatial and temporal separation between the source and receiver of information.

Consistent with the remarks made in Chap. 2 above, the theory becomes a theory of causal actions by the observer on the world when the perspective is switched to that of the environment.

The Barwise-Seligman theory developed in [25] can be seen as fulfilling four desiderata [34] (cf. [35, 36]):

(1) Information flow results from regularities in a distributed system.
(2) Information flow crucially involves both general information about classes of entities, properties, or events, encoded by "types" as well as information about their particulars, encoded by "tokens".
(3) It is by virtue of regularities among connections in the distributed system that information about one component provides information about other components.
(4) The regularities of a given distributed system are enabled by its information channels, and hence can be specified by an analysis of its information channels.

The notion of a *distributed system* clearly evokes segmentation of some overall system into bounded components that pass information back and forth, consistent with the general picture of physical interactions outlined in the previous chapter. Hence we can see the Barwise-Seligman theory as a theory of *physically-implemented* information flow. After outlining the theory in this chapter, we will proceed to discuss its implementation in the chapters that follow.

3.2.1   An Intuitive Example

To build some intuitions for the formalism to follow, suppose your friend Alice is examining a map of London. The map represents "to Alice" a conscious visual percept. At the same time, the map is a conventionalized representation of London. This representation is the result of the semantic relation instantiated in Alice's neurocognitive system, and the semantic relation instantiated by the social conventions of cartography as connected by objects within the map that Alice is examining. Alice's visual experience when examining the map provides information not just about the map, but about London.

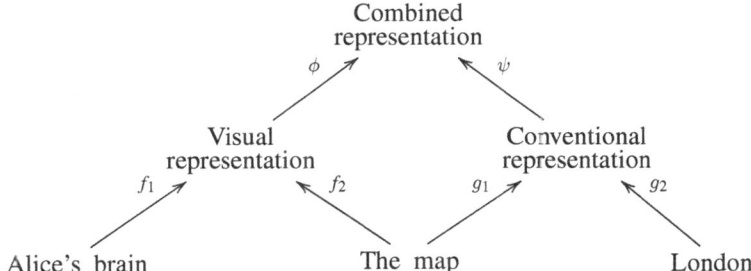

The arrows (formalized below as *logic infomorphisms*) in this diagram represent a bottom-up flow of logical constraints on the meanings of representations.
    The activity pattern in Alice's brain involves visual-processing areas; hence, her representation of the map is constrained (arrow f_1) to be visual. The shapes depicted on the map constrain (f_2) the "shape" that Alice imaginatively experiences.

The shapes depicted on the map likewise constrain (g_1) the geographical locations that it can represent while remaining consistent with the conventions of cartography; the "shape" of London itself similarly constrains (g_2) its set of possible cartographically-accurate maps.

The semantics of Alice's visual representation, and the semantics of the conventional cartographic representation, similarly constrain ϕ and ψ, rendering the combined representation simultaneously visual, cartographic, and of London.

This example illustrates several features of the Barwise-Seligman approach. First, it concerns actionability: Alice is interested in the map because it will help her navigate London, and the map makers adopted the conventions that enabled making the map in order to serve this purpose. Hence, features on the map are precisely "differences that makes a difference" [31]; maps leave out details that are irrelevant to navigation. Second, it not only recognizes but assigns roles to the different interpretations constructed by different agents. The information that Alice interprets the map as carrying is different—and probably quantitatively less—than that seen by the map makers, but she also uses the map differently. There is, moreover, no need to appeal to any causal agency other than the map makers' to explain why the map represents London. Third, the approach treats both the relevant state spaces [25, Chap. 8] [37], and causality [38] explicitly, and recognizes that causation always operates within some context [34–36, 39]. Finally, the theory offers a natural association between information useful for behavioral control, including verbal reporting, and the contents of conscious awareness, as called for by Chalmer's *principle of structural coherence* [40]. Further connections between the Barwise-Seligman approach and Shannon's theory, mainly from the (conditional) probabilistic viewpoint, are discussed in [41, 42], cf. [43] and the discussion below.

3.2.2 The Basic Concepts of Channel Theory

Barwise and Seligman [25] formulate their theory of distributed information flow using *Channel Theory*, is an application of the category theory of Chu spaces (see Appendix B.1). Chu spaces are spaces of semantic relations exemplified by 'object—attribute' tables, originally derived from matrix arrays, e.g. database tables, encoding some information of interest [44–47]. Indeed, Channel Theory can be regarded as defining a category **Chan** isomorphic to the category of Chu spaces **Chu(Set**, K) (for short, **Chu**). We will specify the objects and arrows below (see also Appendices Sect. A, and Sect. B.1 for further details). While conceptually straightforward, Channel Theory is surprisingly rich in its concepts, providing a natural representation of conditional probabilities [41, 42] (see also [48]), a formal criterion for operator commutativity and quantum contextuality [49, 50], along with a hierarchical framework for Bayesian inference [50, 51] (for a logico-philosophical perspective on the properties of semantic coherence, and causality in **Chan**, see e.g. [34]). Furthermore, it provides an implementation-independent formal language for writing functional specifications of QRFs in terms of computations.

> **Box 3.3: Goguen's categorical manifesto**
>
> Category theory was developed in the 1940s as a foundational language in which any mathematical system could be characterized [52]. In 1991, Joseph Goguen published his "categorical manifesto" [53] arguing that category theory provides the most general, and most useful, formal language for talking about computation. It is, therefore, a natural formalism for describing distributed information flow.
>
> A *category* $\mathfrak{C}$ consists of collections of *objects*, and *arrows* (i.e. directed relations, or morphisms) between objects, satisfying two conditions: (1) arrows compose associatively, i.e. for objects A, B, C, D, if $f : A \to B$, $g : B \to C$, and $h : C \to D$, then $hgf : A \to D$, and (2) each object has an identity arrow $\mathrm{id}_A : A \to A$. For further properties, see Appendix A, where the background needed to apply the theory in the current context is provided. There are a number of good introductory textbooks (e.g. [54–56]), as well as the extensive resources available at ncatlab.org.

We proceed with defining the principal concepts used in the Barwise-Seligman theory, beginning with a *classifier* (or, *classification*) that relates tokens in some language to types in that language. We can define a classifier as an object in **Chan**, that complies with a *classification* relation $\Vdash$, as follows:

Definition 3.2.1 A *classifier* $\mathcal{A}$ is a triple $\langle Tok(\mathcal{A}), Typ(\mathcal{A}), \models_{\mathcal{A}} \rangle$ where $Tok(\mathcal{A})$ is a set of "tokens", $Typ(\mathcal{A})$ is a set of "types", and $\models_{\mathcal{A}}$ is a classification relating tokens to types.

A motivation for this definition arises from logical semantics from which Types can be regarded as "formulas" and Tokens as "models".

Example

A First Order Language (FOL) L is a classifier, where $Tok(L)$ consists of a set M of certain mathematical structures, $Typ(L)$ are sentences in L, and $M \Vdash \varphi$, if and only if φ is true in the token M. The type set of a token M is the set of all sentences of L true in M, called the *theory* of M [25, Ex 2.2, p.28] (see also Appendix Sect. C.1).

The classification $\models_{\mathcal{A}}$, as a satisfaction relation, can be valued in a set K having no assumed structure (as is the case for such relations in Chu spaces; see Appendix Sect. B.1). In fact, since **Chan** and **Chu** are isomorphic, essentially equivalent, categories, then any construction (whatever the nomenclature in question) as implemented by one, is applicable to the other. The set K can have several interpretations, such as 'alphabet', etc. Arrows (morphisms) between objects in **Chan**, are called *infomorphisms*, defined as follows:

Definition 3.2.2 Consider classifiers $\mathcal{A} = \langle Tok(\mathcal{A}), Typ(\mathcal{A}), \models_{\mathcal{A}} \rangle$ and $\mathcal{B} = \langle Tok(\mathcal{B}), Typ(\mathcal{B}), \models_{\mathcal{B}} \rangle$, an *infomorphism* $f : \mathcal{A} \to \mathcal{B}$ is a pair of maps $\overrightarrow{f} : Tok(\mathcal{B}) \to Tok(\mathcal{A})$ and $\overleftarrow{f} : Typ(\mathcal{A}) \to Typ(\mathcal{B})$ such that $\forall b \in Tok(\mathcal{B})$ and $\forall a \in Typ(\mathcal{A})$, $\overrightarrow{f}(b) \models_{\mathcal{A}} a$ if and only if $b \models_{\mathcal{B}} \overleftarrow{f}(a)$.

This last definition can be represented schematically as the requirement that the following diagram commutes:

$$\begin{array}{ccc} \mathrm{Typ}(\mathcal{A}) & \xrightarrow{\overleftarrow{f}} & \mathrm{Typ}(\mathcal{B}) \\ {\scriptstyle \models_{\mathcal{A}}}\Big\downarrow & & \Big\downarrow{\scriptstyle \models_{\mathcal{B}}} \\ \mathrm{Tok}(\mathcal{A}) & \xleftarrow{\overrightarrow{f}} & \mathrm{Tok}(\mathcal{B}) \end{array} \qquad (3.11)$$

An infomorphism $f : \mathcal{A} \to \mathcal{B}$ is effectively a map relating the semantic constraints imposed by the classification $\models_{\mathcal{A}}$ to those imposed by $\models_{\mathcal{B}}$. This can be viewed as a way of transmitting information from one classifier to another, so that, for instance, "b is type B" can encode, or represent, the information "a is type A". Hence, the classification relations $\Vdash_{\mathcal{A}}$ and $\Vdash_{\mathcal{B}}$ are explicitly regarded as enforcing semantic constraints, not merely syntactic or set-theoretic constraints, thus rendering both classifiers and infomorphisms as intrinsically semantic notions regulating information flow through a channel. In this way, information is not simply reduced to a quantity of bits (as would be the case for Shannon information [2, 3]), but it is rather subject to logical constraints on the *meanings* assigned to the symbols, as imposed by Definition 3.11. Moreover, the concept of an infomorphism as a mapping between classifiers, also provides the basic framework for constructing, in a scale-free fashion, multi-level, quasi-hierarchical classification systems akin to the connections between processing layers in various descriptive models of Machine Learning and neuroscience. Infomorphisms in such models are intrinsically bidirectional; see Sect. 3.3.1 below for examples.

3.2.3   Logic Infomorphisms from Local Logics and Sequents

The means by which channels encode sets of mutual semantic constraints between classifiers is further elaborated when the classifier concept is fortified by that of a *local logic* [25, Chap. 12], the formal details of which we list in Appendix Sect. C.3 (see in particular Definition C.2.1). For now, we offer a brief explanation of the terms: a classifier is extended to a local logic by specifying a subset (possibly a singleton) of tokens satisfying all of the types of some given (regular) *theory*, as exemplified in the above Example of a FOL, that specifies the logical aspects of a given situation (see Definitions C.1.1 and C.1.2). The theory is *regular* in the sense of regulating the structural properties to which the system in question complies. In fact, any classifier determines its own regular theory, and each local logic incorporates its own regular theory (Definition C.2.1). Accordingly, an infomorphism is extendable to a *logic infomorphism* $\mathcal{L}$ that preserves this additional structure (Definition C.3.1). Given an information flow channel:

$$\longrightarrow \mathcal{A}_{\alpha-1} \longrightarrow \mathcal{A}_{\alpha} \longrightarrow \mathcal{A}_{\alpha+1} \longrightarrow \cdots \qquad (3.12)$$

the semantic and theory content of local logics induces a flow of logic infomorphisms:

$$\cdots \longrightarrow \mathcal{L}_{\alpha-1} \longrightarrow \mathcal{L}_\alpha \longrightarrow \mathcal{L}_{\alpha+1} \longrightarrow \cdots \tag{3.13}$$

Instrumental in the notions of theories and local logics, is that of a *sequent* which, for later purposes, we define now in terms of an 'implication' relation between subsets of types:

Definition 3.2.3 Two subsets $M, N \subseteq Typ(\mathcal{A})$ are related by a *sequent* $M \models_{\mathcal{A}} N$ if $\forall x \in Tok(\mathcal{A}), x \models_{\mathcal{A}} M \Rightarrow x \models_{\mathcal{A}} N$.

Via the sequent relation, the local logic $\mathcal{L}$ assigned to $\mathcal{A}$ effectively arranges the types of $\mathcal{A}$ into a hierarchy in which any classifier can be extended to a local logic by adding types as is necessary. Thus, a sequent encodes a semantic, causal constraint such that in the information flow, it becomes a component of some logical gating mechanism.

Intuitively, we can think of a local logic identifying the token(s) satisfying all of the types, whereby the logic infomorphisms are those transformations that transfer token-identification information between local logics, following which the channels (see below) comprise sets of logic infomorphisms encoding mutual constraints that assemble multiple identified tokens. Collectively, this provides a notion of "logical gating" to the information flow. In [57] we showed that the latter are effectively parts that can be fitted into a larger computational system—effectively, a decision tree—capable of identifying a complex object by identifying both its parts and any required relations between them.

3.3 Cone-Cocone Diagrams

We are particularly interested in collections of infomorphisms that construct complex semantic constraints out of simple ones as these will allow us to specify QRFs as hierarchies of semantic constraints. Given a finite collection $\mathcal{A}_i$ of classifiers, this construction process can be represented diagrammatically by a finite, commuting *Cocone diagram* (CCD) as depicting a flow of infomorphisms sending inputs destined to a *core* $\mathbf{C}'$ that is the category-theoretic *colimit* of the underlying classifiers (see Appendix A.2). Diagrammatically, it is the apex of the maximally general diagram of this form over the $\mathcal{A}_i$ (suitably indexed), providing a unique such maximum exists:

$$
\begin{array}{c}
\mathbf{C}' \\
\overset{f_1}{\nearrow} \quad \uparrow f_2 \quad \overset{f_k}{\nwarrow} \\
\mathcal{A}_1 \xrightarrow{\; g_{12} \;} \mathcal{A}_2 \xrightarrow{\; g_{23} \;} \cdots \mathcal{A}_k
\end{array}
\tag{3.14}
$$

We point out that the cocone core $\mathbf{C}'$ is itself a classifier that encodes, via the incoming infomorphisms f_i, the conjunction of the semantic constraints imposed by the $\mathcal{A}_i$.

3.3.1 Information Flow in Cone-Cocone Diagrams

There is a dual construction to a CCD, namely a commuting finite *cone diagram* (CD) of infomorphisms on the same classifiers, where all arrows are reversed.[1] In this case, the core of the (dual) channel is the category-theoretic limit of all possible downward-going structure-preserving maps to the classifiers $\mathcal{A}_i$. This leads to defining the key concept of a finite, commuting *Cone-Cocone diagram* (CCCD) as consisting of both a cone and a cocone on a single finite set of classifiers $\mathcal{A}_i$ by infomorphisms as depicted below:

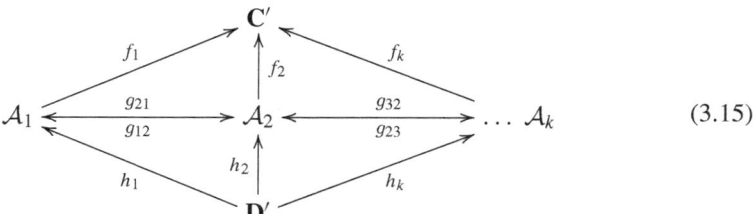

$$(3.15)$$

If the cores $\mathbf{C}' = \mathbf{D}'$, we can also represent the CCCD as:

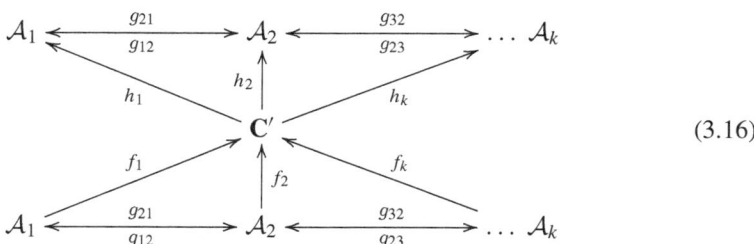

$$(3.16)$$

This diagram is naturally interpreted as reconstructing the semantics of the $\mathcal{A}_i$ via the 'combined' representation $\mathbf{C}'$ and $\mathbf{D}'$. Generalizing Diagram (3.16) by taking $\mathbf{C}'$ be the limit of a smaller set of classifiers $\mathcal{A}'_1, \ldots \mathcal{A}'_j$, $j < k$, leads to:

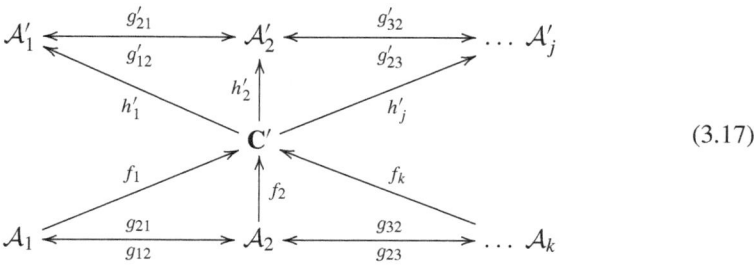

$$(3.17)$$

---

[1] The cocone/colimit construction we outline in Appendix A.2. For the formal descriptions of limits and colimits in a category, see e.g. [55, Sects. 5.5–5.6].

Diagram (3.17) provides a natural representation of coarse-graining the semantics of $\mathcal{A}_i$ via $\mathbf{C}'$ into a compressed representation $\mathcal{A}'_i$. The logic infomorphisms within these diagrams, besides semantically-regulating information flow, also induce embeddings of type hierarchies, such that Diagrams (3.16) and (3.17), can be viewed as embeddings into a "top-level" type hierarchy $\mathbf{C}'$ that assigns an overall semantics to its inputs, followed by encodings of this top-level hierarchy into some componential representation.

Diagrams such as (3.14)–(3.17) can be generalized into layered hierarchical networks by adding intermediate layers of classifiers and the appropriate maps, such that each layer can be seen as a *metaprocessor* complete with an internal memory and read-write system for layers beneath it. Each layer of the overall architecture is, in other words, a functional regulator of the regulators implemented by the lower level processes. How this is relevant to metacognition and its evolution in basal biological systems is discussed in Chap. 8 below.

With these properties, the CCCDs clearly resemble artificial neural networks (ANNs), and in the "bowtie" form of Diagrams (3.16) and (3.17), they represent a basic blueprint for variational autoencoders (VAEs) [58], as earlier formulated in [49]. The core $\mathbf{C}'$ in (3.16), and in (3.17), can be viewed as both an "answer" computed by the f_i from inputs to the $\mathcal{A}_i$ and, dually, as an "instruction" propagated by the h_i (or in Diagram (3.17), the h'_i), to drive outputs from the $\mathcal{A}_i$ (or in Diagram (3.17), the $\mathcal{A}'_i$). Such dual I/O behavior is exactly that of a QRF. We can, therefore, represent any QRF as a CCCD "attached" to a subset of measurement operators $M_k^A, \ldots M_n^A$ by maps that identify the binary eigenvalues of the M_i^A with binary inputs to the $\mathcal{A}_i$ as illustrated in Chaps. 2, Fig. 2.3. This category-theoretic representation is particularly useful in the present context, as it is completely independent of the physical implementation of the relevant QRF. In summary, a CCCD comprises a scale-free architectural model for a massively parallel, distributed information processing system incorporating logical consistency, besides incorporating a process of "reading-from", and "writing-to" some external system [49–51].

Box 3.4: From information flow back to causation

Causality (or causation) affords a rich philosophical background which can be traced back to Plato, Hume, Hobbes, Russell, and others. As noted earlier, Dretske proposed defining information—or more precisely, the semantic content of information—via causal connections between an observer and the world.

John Collier [34], following [59–61], takes the opposite tack, defining causality as the transfer of a particular token of a quantity of information from one state of a system to another. In particular, *physical causality* occurs when physically-encoded information is transferred between states. This notion is consistent with the usual usage of "causality" by physicists, which associates causality with "signaling", i.e. the transfer of information from one system to another at no more than the speed of light.

Collier pursues two principles: (1) Schrödinger's Negentropy Principle of Information (NPI) [62], following which causality is representable by a computational process dynamically embedded in matter or some other framework, and (2) avoiding circularity by defining causality in terms of tokens of information. This is particularly the case for the substantive (i.e. physically encoded) information that permeates biological systems, as reviewed in [63]. For Collier, it is the quantification of form through complexity theory that reveals this information. This leads to an initial definition of a *causal process* [63]:

P is a causal process in system S from time t_0 to t_1, if and only if some particular part of the information of S involved in stages of P is preserved from t_0 to t_1.

Philosophers may cry out for some sort of refinement, which Collier proceeds to do:

P is a causal process in system S from time t_0 to t_1, if and only if some particular part of the form of S involved in stages of P is identical at t_0 and t_1.

Then we need to sharpen up the notion of information transfer:

Information I is transferred from t_0 to t_1 if and only if the same (particular) information exists at t_0 and t_1.

This leads to a further refinement:

P is a causal process in system S from time t_0 to t_1, if and only if some part of the form of S involved in stages of P is transferred from t_0 to t_1.

A straightforward definition of *interactive causation* thus follows:

F is a causal interaction between S_1 and S_2 if and only if F involves the transfer of information from S_1 to S_2, and/or vice-versa.

Note from the previous chapter that for physical systems, "or" is not an option: all interactions are bidirectional, whether the bidirectionality is noticed by the interacting parties or not.

Central to discussions of central cause and common asymmetry are *causal forks* [60] which [63] formulates as:

F is an interactive fork if and only if F is a causal interaction, and F has distinct past branches and distinct future branches.

and

F is a conjunctive fork if and only if F is a causal interaction, and has one distinct past branch and multiple distinct future branches, or vice/versa.

Thus we can see interactive forks as X-shaped, as being open to the future and the past, whereas conjunctive forks are seen as Y-shaped, as being open in only one temporal direction. Overall, *substantive information* (e.g. it from bit, negentropy, hierarchical information, and linguistic information), as discussed in [77], reveals the tight connection between causality and information.

We recall that a distributed system of information flow as above, of which the CCCDs (when considered as hierarchical logical gating systems) embodying context and causation, provide a form of computation via the logic regulation between classifiers [25, 34]. More specifically, for any pair of classifiers $\mathcal{A}$, $\mathcal{B}$ occurring in such a system, there exists some logic infomorphism between them such that $\mathcal{A}$ directly *causes* $\mathcal{B}$. A more abstract, though closely related, theory of integrative information involving logical structures known as *Institutions* [64–67], we will survey in Sect. 3.4 below.

3.3.2 Probability and Bayes' Theorem

Among previous studies connecting logic and information in probabilistic terms are those of Barwise and Perry [36], and Barwise [35]. To see how probabilities can be incorporated into our CCCDs, we commence by returning to the notion of sequent given in Definition 3.2.3. As shown in [41, 42], a probability can be assigned to a sequent by first removing the universal quantifier '$\forall$', and then assigning a probability to $x \Vdash N$, given that $x \Vdash M$, for arbitrary x. Effectively, we assign the probability that x satisfies N, given that it satisfies M. This is clearly a conditional probability that motivates defining:

$$M \Vdash_{\mathcal{A}}^{P} N := P(M|N) \tag{3.18}$$

As noted in [41, 42], this follows from how a conditional probability can be used for interpreting the logical implication "$\Rightarrow$" (see [68]). Given the sequent's conditional probability is p, we then have $M \Vdash_{\mathcal{A}}^{P} N$, observing that it is necessary to have $x \Vdash_{\mathcal{A}} M$, in order to apply the relation $M \Vdash_{\mathcal{A}} N$ in some argument. If the probability of the former holding in $\mathcal{A}$ is $P(M)$, then $x \Vdash_{\mathcal{A}} N$ follows from the usual rule $P(M) \cdot M \Vdash_{\mathcal{A}}^{P} N$. The upshot is that $\Vdash$ is a pre-existing concept, and when interpreted probabilistically it becomes a conditional probability (cf. [48]).

We now see how to construct within a CCCD a hierarchical system of conditional probabilities semantically regulated and preserved by logic infomorphisms, so in this way the CCCD naturally exhibits a probabilistic distribution in a network form. Traversing upwards in the hierarchy imposes multiple conditioning on each "low-level" probability distribution, while traversing downwards sequentially unpacks this conditioning. To realize this in fundamental Bayesian terms, note that M above can be regarded as a previous event, whether it is observed or conjectured, and N as a currently observed datum. Accordingly, $P(M)$ becomes the *prior*, and $P(N)$ the *evidence*, that together generate a prediction. Given the likelihood $P(N|M)$ as the conditional obtained from weakening the sequent (3.18), Bayes' theorem specifies this conditional as the *posterior*:

$$P(M|N) = \frac{P(N|M)P(M)}{P(N)} \tag{3.19}$$

Crucially, the flow of information through a system of logic infomorphisms within the CCCD incorporates the meaning of *inference*. Since a typical classification $\Vdash$ can be considered, not only as causal, but as time- and context-dependent also, these inferential processes can be regarded as having similar dependencies as inferential theory projects. Within the above architecture of the CCCDs, a portion of a typical CCD component for computing hierarchical Bayesian inference from a set of (posterior) observations $\mathcal{A}_i$ to an outcome $\mathbf{C}'$, shapes up as:

$$
\begin{array}{c}
\mathbf{C}' \\
\underset{p_{10}(\cdot|\cdot)}{\nearrow} \uparrow_{p_{20}(\cdot|\cdot)} \overset{p_{k0}(\cdot|\cdot)}{\nwarrow} \\
\mathcal{A}_1 \xrightarrow{\ p_{12}(\cdot|\cdot)\ } \mathcal{A}_2 \xrightarrow{\ p_{23}(\cdot|\cdot)\ } \dots \mathcal{A}_k
\end{array}
\tag{3.20}
$$

We will see in the next chapter how hierarchical CCCDs implement the classical FEP introduced in Chap. 2.

3.4　Institutions

There are certain variations on a theme, and certain generalizations as far the theory of classifications goes (see e.g. [69]). We will proceed to describe one significant generalization for which the theory of classifiers and infomorphisms of [25] provides important essential resources, namely the theory of *Institutions*. Within the development of ideas of Goguen and others [53, 64–66, 70] (see also [67, 71] and cf. [72]), the information flow-classification category (**IFC**) of [25], as outlined in Sect. 3.2.2, is an example of a 1-*Institution*. Generally, the theory of Institutions was originally designed for programming languages implementing a range of abstract logics. Institutions apply category-theoretic methods to formalize an overall sense of 'context' as can be further applied to a wide range of knowledge representations. For instance, they unify the architectural structures of **Chu** and **Chan**, as well as those of Mental and Concept spaces [73, 74], Formal Concept Analysis [75], besides a range of allied logical concepts [64–66, 70]. As pointed out in [71], an Institution comprises an indexing language context, together with a classification passage into the ambient context of classifications. Here, languages or vocabularies are taken as indexing objects for an Institution, in which index linkages are morphisms of such. From the logical perspective, examples of Institutions include first order, intuitionistic, and a range of modal logics [64, 71, 72]. In particular, a focus of the institutional operations concerns *sentences* and *models* as these are parametrized by objects called *signatures* (or *context*) and were originally conceived as a syntactic declaration of metadata. Together with the morphisms (arrows) called *signature morphisms*, this leads to a category **Sign**, from which a set of signatures over a common signature induces a *theory*.

3.4.1 The Basic Concepts

We proceed to specify the above concepts following [64, 65]:

Definition 3.4.1 An *Institution* $\mathfrak{I}$ consists of

(1) a category **Sign** whose objects are called *signatures*;

(2) a functor *Sen* : **Sign** $\longrightarrow$ **Set** giving for each signature a set whose elements are called *sentences* over that signature;

(3) a functor **Mod** : **Sign** $\longrightarrow$ **Cat**^{op} giving for each signature Σ, a category whose objects are called Σ-*models*, whose arrows are called Σ-(model) *morphisms*, and

(4) a relation $\Vdash_\Sigma \subseteq |\mathbf{Mod}(\Sigma)| \times Sen(\Sigma)$ for each $\Sigma \in |\mathbf{Sign}|$, called Σ-*satisfaction*, such that for each morphism $\phi : \Sigma \longrightarrow \Sigma'$ in *Sign*, the *satisfaction condition* $m' \Vdash_{\Sigma'} Sen(\phi)(s)$, if and only if $\mathbf{Mod}(\phi)(m') \Vdash_\Sigma s$, holds for each $m' \in \mathbf{Mod}(\Sigma')$, and for each sentence $s \in Sen(\Sigma)$.

(Note that for any category $\mathfrak{C}$, the notation $|\mathfrak{C}|$ denotes the class of objects of $\mathfrak{C}$). These definitions can be schematically represented by [64, 65]:

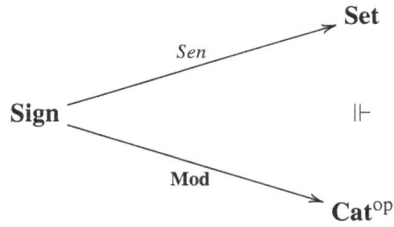

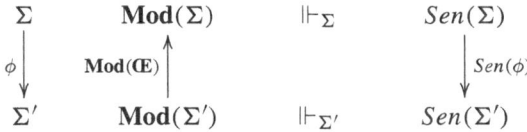

3.4.2 Parametrized Relations

As pointed out in [64], it is sometimes convenient to replace **Mod** : **Sign** $\longrightarrow$ **Cat**^{op}, by a functor *Mod* : **Sign**^{op} $\longrightarrow$ **Set** where *Mod*(Σ) denotes the collection of all Σ-models.

These two versions are related via the equation $Mod(\Sigma) = |\mathbf{Mod}(\Sigma)|$, and entails some slight adjustments in the reading [65]. Given a signature Σ, a parametrization of sentences results via a set $Sen(\Sigma)$ of sentences for each Σ, and for each signature morphism $f : \Sigma \longrightarrow \Sigma'$, a translation $Sen(f) : Sen(\Sigma) \longrightarrow Sen(\Sigma')$.

At the same time, there is a parametrization of models by signatures via an assignment of a class $Mod(\Sigma)$ of models for each Σ, and a translation $Mod(\Sigma') \longrightarrow Mod(\Sigma)$, for each f as before. Sen and Mod are both functors, with the latter contravariant (see above). Then *satisfaction* is a parametrized relation $\Vdash_\Sigma$ pairing $Mod(\Sigma)$ and $Sen(\Sigma)$, such that the satisfaction relation holds: for any signature morphism $f : \Sigma \longrightarrow \Sigma'$, any Σ'-model M', and any Σ-sentence s, we have

$$M' \Vdash_{\Sigma'} f(s) \iff f(M') \Vdash_\Sigma s \tag{3.21}$$

noting here that $f(s)$ abbreviates $Sen(f)(s)$, and $f(M')$ abbreviates $Mod(f)(M')$ (this condition simply expresses 'truth' as a concept invariant under notation).

3.4.3 The Category of Institutions

Original results employing Institutions involve a duality between theories and model classes (typically, we have a Σ-*theory* as a set of Σ-sentences, and a Σ-*model class* is a class of Σ-models, etc.).

Definition 3.4.2 An *Institution morphism* from an Institution $\mathfrak{I}$ to another Institution $\mathfrak{I}'$ consists of a functor $\Phi : \mathbf{Sign} \longrightarrow \mathbf{Sign}'$, together with two natural transformations $a : \Phi; Sen' \longrightarrow Sen$ and $b : Mod \longrightarrow \Phi; Mod'$, such that for any Σ-model M and $\Phi(\Sigma)$-sentence s', we have:

$$M \Vdash_\Sigma a_\Sigma(s') \text{ iff } b_\Sigma(M) \Vdash'_{\Phi(\Sigma)} s' \tag{3.22}$$

Definition 3.4.2 thus specifies a category **INS** with Institutions as objects and morphisms as defined above. A 'dual' notion of an *Institution comorphism* is given in [65, Def 5] leading to a companion category co**INS**, for which both categories are exemplified in fields such as computer algebra with associated logics, formal concept analysis, information flow and cognitive semantics.

3.4.4 Channel Theory and Institutions

Returning to the **IFC** of Sect. 3.2.2, we recall that a classifier (classification) $\mathcal{A}$ is a **1**-*Institution* when $\mathbf{Sign} = \mathbf{1}$, the category with a single object $\bullet$, and with just one morphism. As discussed in [65], **IFC** as a category of 1-Institutions is both generic and important, since it generalizes to arbitrary institutions. In the absence of further generalizations at this stage in time, the category **IFC** is, as we have discussed it, especially suited the QRF construction in assimilating binary qubit strings (see Chap. 2). In this generic case, $Mod(\bullet)$ is the set of Tokens in $\mathcal{A}$, and $Sen(\bullet)$ is the set of Types in $\mathcal{A}$ (this association can be dualized), and institution comorphisms are infomorphisms; thus **IFC** (having limits and colimits of all (small) diagrams [65, Th 3]) can be viewed as a subcategory of co**INS**. Alternatively, if Tokens are taken to be sentences, and Types as models, then **IFC** can be viewed as a subcategory of **INS**.

3.4.5   Local Logics and Institutions

Connections between the categories **IFC** and **INS**, particularly from the perspective of logic classes, are surveyed in [65, 67, 71], again revealing how **IFC** and **INS** are intimately tied. Kent [67] establishes the theory of Institutions to be an indexed, or parametric theory of information flow with classification relations as in Sect. 3.2.2, together with the concepts of local logics and 'theory' [25] (see Appendix B.3), whereby each Institution is covered by a distributed information flow. Crucially, the local logics admitted by **Chan** are the composite logics of an institutional logic environment for which a connecting link is provided by a logic infomorphism (of classifiers) that is both an instance, and a theory link for this environment [67] (see also [71]). All of this can be placed on a grander scale in terms of the concept of a *topos* [78] (we describe the meaning of 'topos' in Appendix Sect. D.2).

We can make the following observation. If a CCCD, as described in Sect. 3.3, is assumed as a model for distributed information flow for some Institution that supports a logical environment, then noncommutativity implies that at least one logic infomorphism up to the core (colimit), is undefinable. Consequently, with regards to Definition 3.4.2, at least one signature (hence one sentence) and satisfaction relation $\Vdash$, is undefinable. This would suggest some possible malfunction in a logical environment leading to some higher order concept of contextuality (or incompleteness/undecidability) occurring within the framework of Institutions, manifestly as between sentences, model types, language, and observation (cf. the quantum treatment of Institutions in [76]).

References

1. Boltzmann, L. *Lectures on Gas Theory*. Dover, New York NY, 1995.
2. Shannon, C. E. and Weaver, W. *The Mathematical Theory of Communication*. Univ. Illinos Press, Urbana IL, 1949.
3. Cover, T. M. and Thomas, J. A. (2006). *Elements of Information Theory*. Wiley, New York.
4. Ashby, W.R. Introduction to Cybernetics; Chapman and Hall: London, UK, 1956.
5. Newell, A. and Simon, H. A. GPS, A program that simulates human thought. In Lernende Automaten, ed. Heinz Billing. Munich: R. Oldenbourg, 1961, pp 109-124.
6. McCulloch, W. S. and Pitts, W. (1943) A logical calculus of the ideas immanent in nervous activity. Bulletin of Mathematical Biophysics. 5(4), 115-133.

7. Gamow, G., Rich, A. and Yčas, M. The problem of information transfer from the nucleic acids to proteins. In Advances in Biological and Medical Physics, (J. Lawrence and C. Tobias, editors), New York, Academic Press, Inc., 1956, pp. 23-68.
8. Feynman, R. P., Leighton, R. B. and Sands, M. (1965). *Feynman Lectures on Physics*. New Yorlk: Addison-Wesley.
9. Wheeler, J.A. Information, physics, quantum: The search for links. In Complexity, Entropy, and the Physics of Information; Zurek, W.H., Ed.; Westview: Boulder, CO, USA, 1990; pp. 3–28.
10. Landauer, R. Irreversibility and heat generation in the computing process. IBM J. Res. Dev. 1961, 5, 183-195.
11. Bennett, C.H. The thermodynamics of computation. Int. J. Theor. Phys. 1982, 21, 905-940.
12. Bennett, C. H. Notes on Landauer's principle, reversible computation, and Maxwell's Demon. Studies in the History and Philosophy of Modern Physics 34 2003, 501–510.
13. Parrondo, J.M.R., Horowitz, J.M. and Sagawa, T. Thermodynamics of information. Nat. Phys. 2015, 11, 131-139.
14. Parrondo, J. M. Thermodynamics of information. 2023. arXiv:2306.12447v1 [cond-mat-stat-mech]
15. Lyon, A. Philosophy of probability. In Allhoff, F. (Ed) Philosophies of the Sciences. Blackwell, London, 2010, pp. 92-125.
16. Peterson, M. Philosophy of probability. International Encyclopedia of Statistical Science, 2014, pp. 1065-1069.
17. Hájek, A. Interpretations of probability. Stanford Encyclopedia of Philosophy, 2023 (https://plato.stanford.edu/entries/probability-interpret/).
18. Mermin, D. Making better sense of quantum mechanics. Reports on Progress in Physics 2017; 82(1), 012002.
19. Bormashshenko, E. The Landauer principle: reformulation of the second thermodynamics law or a step to great unification? *Entropy* 21 2019, 918.
20. Street, S. Upper limit on the therodynamic information content of an action potential. Front. Syst. Neurosci. 14 Article 37, 1–7.
21. Vaccaro, J. A. and Barnett, R. M. Information erasure without an energy coast. *Proc. R. Soc, A* 467 2011, 1770–1778.
22. Floridi, L. *The Logic of Information: A Theory of Philosophy as Conceptual Design*. Oxford Univ. Press, 2019.
23. Dretske, F. *Knowledge and the Flow of Information*. Cambridge MA: MIT Press, 1981.
24. Lombardi, O. Dretske, Shannon's theory and the interpretation of information. *Synthese* 144 2005, 23–39.
25. Barwise, J. and Seligman, J. (1997). *Information Flow: The Logic of Distributed Systems* (Cambridge Tracts in Theoretical Computer Science, 44). Cambridge, UK: Cambridge University Press.
26. Gödel, K. Über formal unentscheidbare sätze der Principia Mathematica und verwandter systeme, I, *Monatsh. Math. Phys.* 1931; 38(1),
27. Wittgenstein, L. Philosophical Investigations. London, Blackwell, 1953.
28. Chang, T. A. and Bergen, B. K. (2024) Language model behavior: A comprehensive survey. Comput. Ling. 50(1), 293-350.
29. Mitchell, M. and Krakauer, D. C. (2023). The debate over understanding in AI's large language models. Proc. Natl. Acad. Sci. USA 120, e2215907120.
30. Kripke, S. Naming and Necessity. Cambridge, MA, Harvard University Press, 1980.
31. Bateson, G. *Steps to an Ecology of Mind: Collected Essays in Anthropology, Psychiatry, Evolution, and Epistemology*. Jason Aronson, Northvale, NJ, USA, 1972.
32. Maturana, H. R. and Varela, F. J. (1980). Autopoiesis and cognition: The realization of the living. Reidel, Boston.

33. Roederer, J. (2005). *Information and its Role in Nature*. Heidelberg: Springer.
34. Collier, J. (2011). Information, causation and computation. In (G. D. Crnkovic and M. Burgin, eds.) *Information and Computation: Essays on Scientific and Philosophical Foundations of Information and Computation*, pp. 89-105. *World Scientific Series in Information Studies Vol 2*. World Scientific Press. Hackensack, NJ.
35. Barwise, J. (1997). Information and Impossibilities. *Notre Dame Journal of Formal Logic* 38(4), 488–515.
36. Barwise, J. and Perry, J. (1983). *Situations and Attitudes*. Cambridge, MA: Bradford Books, MIT Press.
37. Barwise, J. State spaces, local logics, and non-monotinicity. *Logic, language and computation, Vol 2. (London, 1996)*, pp 1–20, *CSLI Lecture Notes* **96**, CSLI Publ., Stanford, CA, 1999.
38. Pearl, J. *Causality*. Cambridge Univ. Press, Cambridge UK, 2009.
39. Seligman, J. Situation Theory Reconsidered. In (Baltag, A. and Smets, S.) Johan van Benthem on Logic and Information Dynamics. Outstanding Contributions to Logic, vol. 5, pp. 895–932. Springer, Cham, 2014.
40. Chalmers, D. J. Facing up to the problem of consciousness. *J. Consciousness Studies* 2(3) (1995), 200–219.
41. Allwein, G. A qualititative framework for Shannon Information theories. In *NSPW '04: Proceedings of 2004 Workshop on New Security Paradigms* 2004 (Nova Scotia, Canada, September 20-23, 2004). New York: ACM (pp. 23-31).
42. Allwein, G, Yang, Y. and Harrison, W. L. Qualitative decision theory via Channel Theory. *Logic and Logical Philosophy* 20 2011, 1–30.
43. Seligman, J. (2009). Channels: From logic to probability. In: Sommaruga, G. (Ed.) *Formal Theories of Information. Lecture Notes in Computer Science* 5363 Berlin: Springer (pp. 193–233).
44. Barr, M. (1979). *-Autonomous categories, with an appendix by Po Hsiang Chu. *Lecture Notes in Mathematics* 752. Berlin: Springer.
45. Pratt, V. (1995). Chu spaces and their interpretation as concurrent objects. *Lecture Notes in Computer Science* 1000, 392–405.
46. Pratt, V. (1997). Types as Processes, via Chu spaces. Invited paper. Proceedings 'Express'97: Fourth Workshop on Expressiveness in Concurrency'. Santa Margherita, Italy. September 1997. *Electronic Notes in Theoretical Computer Science* 7 21pp. http://www.elsevier.nl/locate/entcs/volume7.html
47. Pratt, V. (1999). Chu spaces. *School on Category Theory and Applications (Coimbra 1999)*, Vol. 21 of *Textos Mat. Sér. B*, University of Coimbra, Coimbra. (pp. 39–100).
48. Wobcke, W. An information-based theory of conditionals. *Notre Dame J. Formal Logic* 41(2) (2000), 95–140.
49. Fields, C. and Glazebrook, J. F. (2019a). A mosaic of Chu spaces and Channel Theory I: Category-theoretic concepts and tools. *Journal of Experimental and Theoretical Artificial Intelligence* 31(2), 177-213.
50. Fields, C. and Glazebrook, J. F. Information flow in context-dependent hierarchical Bayesian inference. *J. Expt. Theor. Artif. Intell.* 34 2022, 111–142.
51. Fields, C., Friston, K.J., Glazebrook J.F. and Levin M. A free energy principle for generic quantum systems. Prog. Biophys. Mol. Biol. 2022, 173, 36–59.
52. Eilenberg, S. and Mac Lane, S. (1945). Relations between homology and homotopy groups. *Ann. Math.* 46, 1945, 480–509.
53. Goguen, J. A. (1991). A categorical manifesto. *Mathematical Structures in Computer Science* 1, 49–67.

54. Adámek, J., Herrlich, H. and Strecker, G. E. *Abstract and Concrete Categories: The Joy of Cats*. Wiley, New York, 2004. Available at http://katmat.math.uni-bremen.de/acc (Accessed May 26, 2019).

55. Awodey, S. *Category Theory*. (Oxford Logic Guides, 62). Oxford, UK: Oxford University Press, 2010.

56. Fong, B. and Spivak, D. Seven Sketches in Compositionality: An Invitation to Applied Category Theory. arxiv:1803.05316, 2018.

57. Fields, C and Glazebrook, J. F. A mosaic of Chu spaces and Channel Theory II: Applications to object identification and mereological complexity. *Journal of Experimental and Theoretical Artificial Intelligence* 31(2) 2019, 237–265.

58. Kingma, D. P and Welling, M. An Introduction to variational autoencoders. *Foundations and Trends in Machine Learning* 12(4) (2019), 307–392.

59. Dowe, P. Wesley Salmon's Process Theory of Causality and the Conserved Quantity Theory. *Philosophy of Science* 59 (2001), 195–216.

60. Salmon, W. C. *Scientific Explanation and the Causal Structure of the World*. Princeton Univ. Press, Princeton NY, 1984.

61. Salmon, W. C. Causality without counterfactuals. *Philosophy of Science* 61 (1994), 297–312.

62. Schrödinger, E. *What is Life?* (1944). Reprinted in *What is Life? And Mind and Matter*. Cambridge Univ. Press, Cambridge UK, 2006.

63. Collier, J. Information in biological systems. In (P. Adriaans and J. van Bentham) *Handbook of Philosophy of Science, Vol. 8: Philosophy of Information*, Ch. 5, pp 763–787. North Holland, Dordrecht, 2008.

64. Goguen, J. and Burstall, R. (1992). Institutions: Abstract model theory for specification and programming. *J. Assoc. Comp. Mach., 39*, 95–146.

65. Goguen, J. Information integration in Institutions. 2005 Proposed for: Moss, L. (Ed.) *Thinking Logically: A Memorial Volume for Jon Barwise*. Bloomington IN: Indiana University Press (https://cseweb.ucsd.edu/~goguen/pps/ifi04.pdf).

66. Goguen, J. Three perspectives on information integration. Dagstuhl Seminar Proceeding 04391. Semantic Operability and Integration https://drops.dagstuhl.de/opus/volltexte/2005/38

67. Kent, R. E. The Institutional approach. In (Poli, R., Healey, M. and Kameas, A., eds.) *Theory and Applications of Ontology: Computer Applications*, 533–563. Springer Link, 2010.

68. Adams, E. W. *A Primer of Probabilistic Logic*. University of Chicago Press, Chicago IL, 1988.

69. Parrochia, D. and Neville, P. *Towards a General Theory of Classifications*. Birkhäuser, Basel, 2012.

70. Goguen, J. What is a concept? In (F. Dau and M.-L. Mungier eds.) *Proceedings, 13th Conference on Conceptual Structures*. Lect. Notes in Artificial Intelligence 3596, pp. 52–77. Springer, Kassel, Germany, 2005.

71. Kent, R. E. Semantic Integration in the Information Flow Framework. 2018 arXiv:1810.08236v1 [cs.LO]

72. Spivak, D. and Kent R. E. Ologs: a categorical framework for knowledge representation. *PLOS ONE* 7(1) 2012, e24274.

73. Fauconnier, G. *Mental Spaces: Aspects of Meaning Constructions in Natural Language*. Bradford Books MIT Cambridge MA, 1985.

74. Gärdenfors, P. *Conceptual Spaces: The Geometry of Thought*. Bradford, 2000.

75. Ganter, B. and Wille, R. *Formal Concept Analysis: Mathematical Foundations*. Springer Berlin, 1997.

76. Caleiro, C., Mateus, P., Sernadas, A. and Sernadas, C. Quantum Institutions. In (K. Futasugi et al. eds.) *Goguen Festschrift*. Lecture Notes in Computer Science 4060, pp. 50–64. Springer-Verlag Berlin Heidelberg, 2006.

77. Collier, J. Kinds of information in scientific use. *tripleC (Cognition Communication Co-operation)* 9(2) (2011), 295–304.

78. Kent, R. E. Conceptua: Institutions in a topos. arXiv:1811.02041v1 [cs.LO] (2018).

Hierarchical Bayesian Inference and the Classical FEP

This book is about distributed information flow in generic quantum systems. We have already hinted that the CCCDs defined in the previous chapter can be employed as functional specifications of the QRFs defined in Chap. 2. "Function" is, moreover, a semantic notion, as we saw in Chap. 1. We have, therefore, the necessary ingredients to *express generic physical interactions in semantic terms*, that is, it terms of what they *mean* to the interacting systems. To anyone brought up on the notion that physics is about mechanical interactions between inert objects—"particles" characterized just by mass, charge, and angular momentum—this statement may seem counter-intuitive, or even ludicrous. If, however, one adopts Wheeler's view that physics is about information exchange between "observer-participants", the idea that the information being exchanged might have some meaning becomes obvious: why else would they spend energy to exchange it? Meaning, after all, is actionability, the "difference that makes a difference" in what the receiver *does with* the information received. Why would systems bother to exchange information that was completely useless, that had no effect at all on what they did next?

The paradigm cases of useful information exchange do not, of course, come from physics. They come from psychology, and particularly from human psychology. It was, after all, the need to assure high fidelity in human communications that motivated the development of information theory, and it was the need to replicate and extend human computational abilities that led to the development of computers. Humans are the canonical cognitive species, though the field of animal cognition continually generates surprises [1], and the growing "basal cognition" movement treats all living systems as cognitive [2–4]. Hence before delving deeply into the characterization of generic systems, it is useful to develop some background—and intuitions—by examining cognition in humans and other organisms.

We will focus on two features of human cognitive architecture: its division into "object" and "meta" level processes, and its reliance on prediction. The first of these leads to the

C. Fields and J. Glazebrook, *Distributed Information and Computation in Generic Quantum Systems*, Synthesis Lectures on Engineering, Science, and Technology, https://doi.org/10.1007/978-3-031-97263-8_4

study of metacognitive or "executive" control, resource allocation and attention, shared-memory structures, and the *global workspace* (GW) or global *neuronal* workspace (GNW) theories developed by Bernard Baars [5] and Stanislaus Dehaene [6], and their respective colleagues. The second leads to the reconceptualization of learning as the development of expectations, predictive-coding models and the "Bayesian Brain" hypothesis, and the Free Energy Principle (FEP) introduced in Chap. 2. As we will see, describing information flow with CCCDs helps simplify and unify these two conceptual and theoretical threads.

As befits the literature to which we will refer, we will use classical language throughout this chapter. In the next chapter, we will introduce a problem that confronts any classical analysis of cognition: the problem of *intrinsic contextuality*. This problem is not just conceptual, but is also formal: it leads to—or viewed the other way, is defined by—the joint probability distributions on which the analysis in this chapter depends ceasing to be well-defined. We will diagnose this problem using CCCDs in Chap. 5, and address it by giving the FEP a quantum formulation in Chap. 6. We will, at that point, have a theory of distributed information flow in generic quantum systems.

4.1   Perception-Action Loops, Power, and Control

Recall Landauer's principle: irreversibly writing a bit costs at least $\ln 2 k_B T$. This energetic cost, for any system S, of acting on its environment E is enshrined in the interaction Hamiltonian H_{SE} given in Chap. 2, Eq. (2.2). All information processing systems, therefore, require power to write their outputs, even if they are fully-reversible (e.g. quantum) computers on the inside. This power must be sourced from E in the form of *thermodynamic free energy* (TFE) as discussed in Chap. 2. The fundamental building block for any computer is, therefore, not a central processor or a memory, but rather a power supply. Without externally-sourced TFE, nothing happens.

Once a system has power, it can read information from its environment, perform computations on the information, and write the results back to its environment. Physicists call this process "interaction", computer scientists call it "computation", and psychologists call it a "perception-action loop". Bacterial chemotaxis is one of the best studied examples of a perception-action loop in a living organism; see [7] for an accessible review. Many bacteria are capable of sensing both beneficial (e.g. sugars) and harmful (e.g. acids) molecules in their environments via receptors on their "front" ends, and responding with either forward-directed motion ("running") or random axis re-orientation ("tumbling"). The switch between the two is a simple if-then-else switch: if "good", run; else tumble. The switching mechanism is implemented by phosphorylation of a protein, CheY, via the action of another protein, CheA, that is coupled to the receptors that probe the environment. Phosphorylated CheY binds to a "motor" protein that uses energy derived from ion flow across the cell membrane to spin a flagellum in the "run" direction. The states of a number of other proteins, including

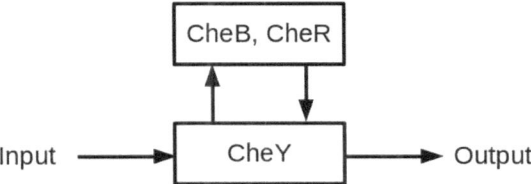

Fig. 4.1 Representation of the *E. coli* chemotaxis pathway [7] as a hierarchical control system. The modulatory proteins CheB and CheR, together with other factors, regulate the sensitivity of the switch implemented by CheY

CheB and CheR, provide a memory that is longer than the switching time by modulating the activity of CheY.

The phosphorylation state of the CheY protein in this simple system implements a binary reference frame: the phosphorylated state indicates a "good" environment and induces approach behavior, while the unphosphorylated state indicated a "bad" environment and induces withdrawal behavior [8]. The states of the modulatory proteins CheB and CheR, together with other, slower-varying properties of the intracellular environment, effectively control the sensitivity of the CheY switch; in Bayesian terms, they control the precision assigned to input from the environment. We can represent this coupled system as a two-level hierarchy, as in Fig. 4.1. The "upper" level process works, at a relatively slower timescale, to control the activity of the "lower" level system, which processes information at a faster timescale. The upper level is, therefore, a *metacontroller* or *metaprocessor* over the lower, *object-level* system. As shown in [9], natural selection can be expected to favor such hierarchical architectures whenever a system is faced with selective pressures acting at two different timescales. Hence it is not surprising to find hierarchical structure in an evolutionary ancient perception-action loop. Indeed, such hierarchical structures are even present in biochemical pathways, e.g. within signal transduction and gene regulatory networks, acting inside biological cells [10, 11].

Box 4.1: 4EA cognition and the "no information processing" claim

One often encounters claims in the literature of Embodied, Embedded, Enactive, Extended, and Affective (4EA) cognition that there are "no representations", "no computations", or even "no information processing" in organisms or other cognitive systems [12–16]. What are we to make of such claims, particularly in view of the idea that a "computation" is just a functional description of a system's physical dynamics [17], as discussed in Chap. 1? Are some advocates of 4EA cognition claiming that the concentration of phosphorylated CheY does not "represent" the concentration of "good" molecules in the environment of *E. coli*, even though this is evidently its functional role in the system? Or that *E. coli*, let alone humans, are doing no information processing whatsoever as they navigate and act on their environments? Apparently so.

A passage from Anthony Chemero provides a good sense of what is going on here: "It is only for convenience (and from habit) that we think of the organism and environment as separate; in fact, they are best thought of as forming just one nondecomposable system. Rather than describing the way external (and internal) factors cause changes in the organism's behavior, such a model [a dynamical systems model] would explain the way the system as a whole unfolds over time" [16] pp. 148–149. Dynamical systems are defined on vector spaces, e.g. classical configuration space. We know from introductory linear algebra, or from Chap. 2, Sect. 2.1 that any vector space can be decomposed. So what does it mean to say that a system is "nondecomposable"? One option would be that the system is in an entangled state, or in classical language, that it is characterized by ubiquitous nonlocal causality; in this case only "the system as a whole" would have well-defined dynamics. We would not be able to "think of the organism and environment as separate" because we would not be able to distinguish them as entities. We would not, in particular, be able to make any predictions at all about how the organism *per se* behaves, as its behavior would be both causally and computationally inseparable from that of its environment.

Note, moreover, that dynamical systems are (collections of) functions, and so a dynamical system model of the observed dynamics of some physical system is a computational model in the sense of [17], and hence in the sense used here. There is nothing "non-computational" about dynamical systems models, as work in reservoir computing [18] makes clear. This remains the case whether a system interacting with an environment or an inseparable joint system is being modeled. Hence 4EA models are computational, even if one adopts an assumption of inseparability as proposed by Chemero.

In the sections that follow, we will generalize only slightly from the case of bacterial chemotaxis, but will find that by doing so we have a model that describes, with only quantitative elaborations, a broad range of information-processing systems, including ourselves. We will, in particular, assume a system S that has a power supply, some finite number of independent, non-interacting perception-action loops (PALs), and a metaprocessor. We assume that the PALs all have finite resolution for both perception and action. We assume that the metaprocessor (MP) has the following capabilities:

1. It has access to the TFE usage of each of the PALs, and controls the flow of TFE to each PAL to either enhance or inhibit its activity.
2. It has access to a read/write memory, one sector of which allows it to read inputs from and write outputs to the PALs.
3. Its memory is large enough that it can store, and hence process, multiple observational outcomes from, and state-preparation inputs to, at least some subset of the PALs.

The first of these gives S a limited self-monitoring capability: the MP can track the TFE usage of the rest of the system and can take steps to selectively limit TFE usage if TFE

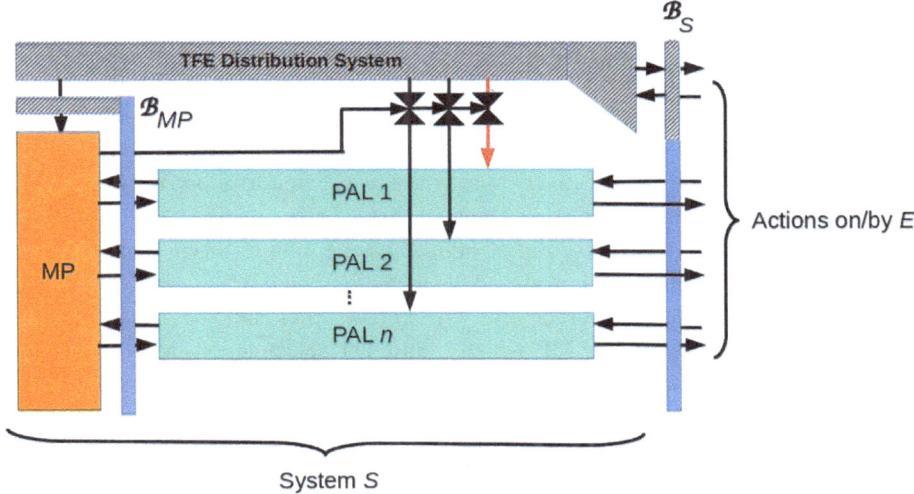

Fig. 4.2 Cartoon representation of the computational architecture assumed here. Hatched components of the interfaces $\mathcal{B}_S$ and $\mathcal{B}_{MP}$ of the system S and its MP, respectively, indicate thermodynamic sectors. Adapted with permission from [19] Fig. 3

supplies are inadequate. As regulating TFE allocation to individual PALs implements a form of attentional control of both perception and action, self-monitoring and attention are closely coupled in this model. The second of these capabilities allows the MP to obtain outcomes from and to control the state-preparation activities of the PALs, and hence to learn from, and act on, S's world via the PALs. Note that the PALs, together with TFE distribution system and the memory, constitute the MP's environment; hence the MP affects S's environment indirectly by acting on its own environment. The third capability gives the MP the key resource—a "scratchpad" or short-term, working memory—needed to execute a planning algorithm that requires data from multiple timesteps of the PALs. The resulting architecture is illustrated in cartoon form in Fig. 4.2.

We will make no assumption at this point about the implementation of either the PALs or the MP, other than that the architecture of S as a whole exemplifies some Turing-complete architecture. Note that this is *not* an assumption that S can compute any computable function; S can only compute functions for which it has the resources, and can only compute such functions with inputs that can be specified as inputs, at the available resolution, to one or more of the PALs. We will discuss a particular architecture in Sect. 4.4 below, and will show in the next chapter that the PALs can always be represented as QRFs.

4.2   Shared Memory and Global-Workspace Architectures

The architecture shown in Fig. 4.2 is a simple form of shared-memory architecture: the PALs cannot access each other's sectors of either $\mathscr{B}_S$ or $\mathscr{B}_{MP}$, but each PAL shares access to a sector of $\mathscr{B}_{MP}$ with the MP. The actions of the PALs can, therefore, influence each other via the MP. Hence $\mathscr{B}_{MP}$ is, effectively, a synchronous shared memory between any PAL and the MP, and an asynchronous shared memory between any two PALs.

Architectures with the general form of Fig. 4.2 have a long history in cognitive neuroscience. In 1983, Bernard Baars introduced the idea of a *global workspace* (GW) in which inputs from exteroceptive and interoceptive sensory modalities could be pooled for analysis by an *executive* or *metacognitive* MP, which would then select actions to be implemented by one or more externally- or internally-directed action modalities [5]. Critically, the GW served as a working memory for the executive MP that allowed it to track the time correlations between distinct sensory inputs and the (short-term) consequences of taking particular actions. The executive MP was identified anatomically with the prefrontal cortex (PFC) in humans, and the GW with connections between the PFC and the evolutionarily-older parts of the brain.

Box 4.2: GW models and the "theater of consciousness"

The GW model was motivated by the problem of understanding human consciousness. Baars described the GW as the "theater of consciousness" in which whatever data were lit up by the "spotlight of attention" were processed and acted upon [20]. Later work by Baars and colleagues [21, 22] and, under the rubric of the global *neuronal* workspace (GNW) theory, by Stanislaus Dehaene and colleagues [6, 23, 24] both elaborated the neuro-anatomical implementation of the G(N)W and its role in conscious perception and action.

Recalling from Chap. 2 the idea that classical data—and hence classical inputs and outputs—are encoded on boundaries, we can read off from Fig. 4.2 the loci of experienceable inputs or "sensations". The system S experiences what is encoded on $\mathscr{B}_S$, while its MP component experiences what is encoded on $\mathscr{B}_{MP}$. The internal boundary $\mathscr{B}_{MP}$ encodes sensory data (for S) that have been processed by S's PALs. If we think of PALs as "packages" of cognitive processes that identify and characterize objects as described in Sect. 2.5.3, we can think of $\mathscr{B}_{MP}$ as encoding identified "objects" as they appear in "scenes". Bernard Hommel has called this the "event file" level of description [25]. It corresponds, up to attentional modulation, with what we ordinarily think of as "perceptual experience". In complex systems such as humans or other animals, this includes perceptual experiences that originate in the body, i.e. interoceptive experience [26]. One functional role of the G(N)W in all models is multi-sensory fusion, including fusion of percepts originating inside and outside of the body [27].

Figure 4.2 does not provide a locus for *imaginative* experience, e.g. mental imagery, inner speech, or "thoughts" in some other modality. Hence it does not include any locus for an experienced encoding of the "self-model" that the MP employs to control the behavior of the PALs, e.g. by differential allocation of attention. As discussed in [19], either of these functions requires an additional "internal" boundary, together with gating functions that allow "copies" of sensory data to be encoded on this additional boundary.

The GW concept has become progressively more sophisticated as both experimental and theoretical methods have developed over the past three decades. The combination of whole-brain functional imaging and network modeling methods has driven the development of neuro-architectural models of both executive control and data sharing across sensory, memory, and motor networks, variously referred to as the *giant component* [28] or the *connective core* [29] of the frontal-parietal-temporal cortical network. The role of deep brain structures in "driving" activity in the GW via modulating arousal and effective connectivity [22] has also become much better defined [30, 31]. The GW has also been explicitly recognized as managing communication between exteroceptive and interoceptive perception and action pathways, and hence as linking the perception and behavior of the body's internal state to perception of and behavior in the external environment as shown in Fig. 4.3 [27].

As a shared-memory architecture, the GW model has been abstracted into AI architectures, more explicitly in LIDA [32]. Any system in which multiple PALs share information via a read/write memory can, in principle, be viewed as implementing a GW architecture.

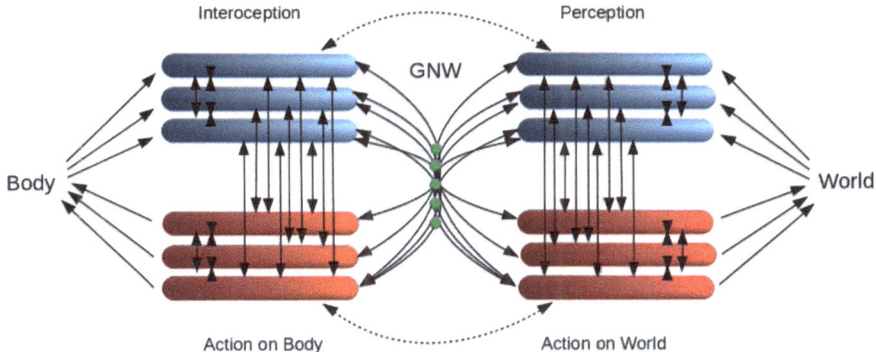

Fig. 4.3 Simplified cartoon of a GNW multi-parallel distributed architecture incorporating both perception of the external world and interoception of bodily state, and both action of the external world (i.e., overt behavior) and regulation of bodily state (e.g., regulation of hormonal signals and blood pressure). Bidirectional vertical arrows indicate non-GNW modulatory connections between input and output processing modules on either the perception or interoception side; green dots represent GNW hubs. Upper and lower dashed arcs connecting perception and action systems are lateral (i.e., cross-modulatory) connections induced by requiring the GNW nodes to be cores of cocones spanning both world- and body-directed systems. These arcs transfer the information needed to generate epistemic feelings as synthetic interoceptions blending cortical with subcortical information. From [27] Fig. 2; used with permission

4.3   Hierarchical Bayesian Inference

Recall from Chap. 3 that Bayes' theorem states that, for events x and y, if the probability $P(y) \neq 0$, then the conditional probability of x given y, i.e. $P(x|y) = [P(y|x)P(x)]/P(y)$ (cf. Eq. (3.19)). This equality follows trivially from the Kolmogorov axioms for probabilities, i.e. that probabilities are non-negative, finite and bounded—so can be normalized to have values in $[0, 1]$—and that probabilities of disjuncts add while probabilities of conjuncts multiply. Hence Bayesian inference is just inference consistent with the usual rules of probability.

As a hierarchy of operations that respect the Kolmogorov axioms, a CCCD implements Bayesian inference. As noted in Chap. 3, fully-connected, feed-forward artificial neural networks (ANNs) are CCCDs. If the connection weights in ANNs are constrained so as to respect the Kolmogorov axioms in their representation as probabilities, then ANNs can be viewed as Bayesian classifiers [33]. Assembling a complex type or entity identification by combining items of low-level evidence, e.g. as illustrated in Fig. 4.4, while respecting the Kolmogorov axioms provides an example; see [34] for further examples and discussion.

Hierarchical Bayesian inference provides the architectural basis for Bayesian brain [35, 36] and predictive coding [37–39] models in neuroscience. The most prominent of these is the theoretical framework provided by the classical free energy princple (FEP) briefly introduced in Box 2.3.

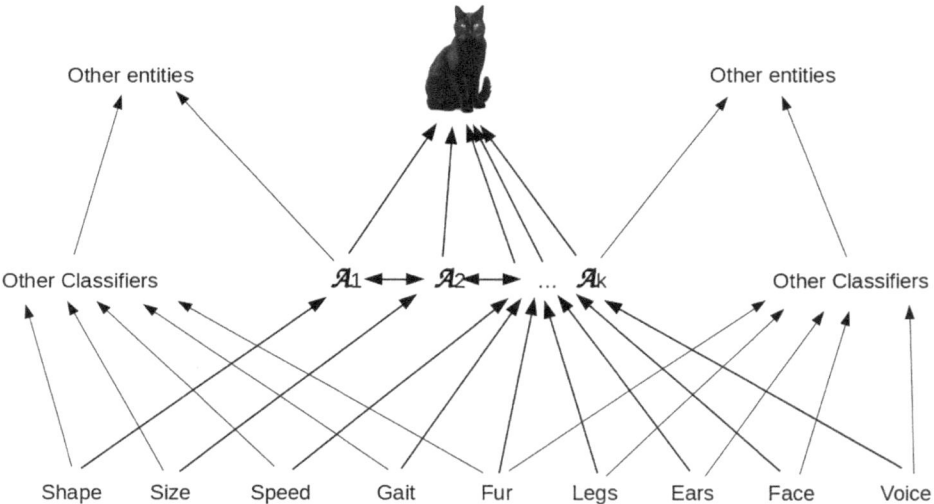

Fig. 4.4 A type or object recognition problem is solved when a cocone above the relevant subset of classifiers exists. Lower-level features (i.e. cores of cones) typically map to many classifiers, which in turn may map to distinct types or individual objects (i.e., cores of cocones). Individual objects can be viewed as single-member types. From [27] Fig. 1; used with permission

4.4   The Classical FEP

The FEP describes dynamical systems that can be factored, for the duration of some time period of interest, into conditionally statistically independent components, which we will refer to as a "system" S and its "environment" E. The FEP states, in obviously tautological form, that in order to persist as separate systems, S and E must both behave in a way that enables their continued statistical independence [40]. A system compliant with the FEP—what Friston [40] calls a "thing" or "particle"—can be regarded as having a boundary composed of a set of states that function as a Markov blanket (MB) [41], i.e. a set of states on which all interactions with the environment E can be conditioned. Formally, a (minimal) MB around a set of states X in a causal network comprises the parents not in X of states of X, the children not in X of states of X, and any parents not in X of children of X that are not in X; Fig. 4.5 illustrates this definition.

Adopting the notation of [40], if the internal states μ (i.e. the states comprising the "internal" system X above) of a system S are conditionally independent of the states η of its environment E, it possesses an MB, defined as the set b of states on which the dependence of μ on η is conditioned. This MB condition will hold, in general, provided the interaction between S and E is significantly weaker than the internal self-interactions of either; this is the condition of "sparse coupling" between S and E. Three remarks are in order here. First, the environment E is the *entire* environment of S, or at least the entire environment of interest; in what follows, we will always consider E to comprise "everything but S." Second, fixing the sets of states μ and η uniquely fixes the set b and hence the MB. Every system, therefore, has a unique MB separating it from its environment. Third, the definition

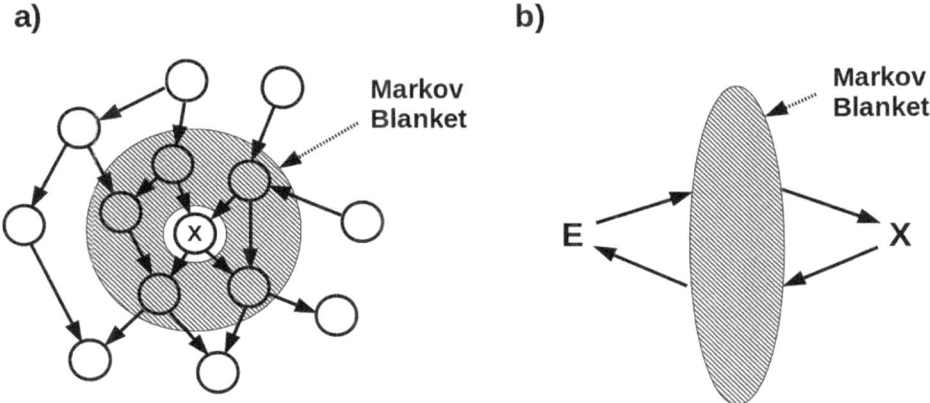

Fig. 4.5 a An MB around a set of states X in a causal network comprises the parents not in X of states of X, the children not in X of states of X, and any parents not in X of children of X that are not in X. **b** An MB around X can be viewed as a boundary separating X from an environment E; together X and its MB comprise a persistent system or "thing" S. From [42] Fig. 2; used with permission

of an MB makes no explicit reference to ordinary, three-dimensional (3d) space. Therefore, while it is commonplace to consider the MB of an organism, for example, to coincide with or be implemented by its 3d spatial boundary, this is not a requirement of the theory. Indeed, nothing in principle prevents a collection of spatially-disconnected entities, e.g. a population of organisms, from having an MB.

Given the conditions above, a variational free energy (VFE) can be defined as a statistical relationship between internal (μ), external (η), and intervening blanket (b) states [43]. This VFE can be written [40], Eq. 2.3:

$$
\begin{aligned}
F(\pi) &= \underbrace{\mathbb{E}_{q(\eta)}[\ln q_\mu(\eta) - \ln p(\eta, b)]}_{\text{Variational free energy}} \\
&= \underbrace{\mathbb{E}_q[-\ln p(b|\eta) - \ln p(\eta)]}_{\text{Energy constraint (likelihood \& prior)}} - \underbrace{\mathbb{E}_q[-\ln q_\mu(\eta)]}_{\text{Entropy}} \\
&= \underbrace{D_{KL}[q_\mu(\eta)|p(\eta)]}_{\text{Complexity}} - \underbrace{\mathbb{E}_q[\ln p(b|\eta)]}_{\text{Accuracy}} \\
&= \underbrace{D_{KL}[q_\mu(\eta)||p(\eta|b)]}_{\text{Divergence}} \underbrace{- \ln p(b)}_{\text{Log evidence}} \geq -\ln p(b)
\end{aligned}
\tag{4.1}
$$

where $\pi(t) = (\mu(t), b(t))$ is the time-dependent "particle" state. The VFE functional $F(\pi)$ is an upper bound on surprisal (a.k.a. self-information) $\mathfrak{I}(\pi) = -\log P(\pi) = -\ln p(b)$ because the Kullback-Leibler divergence term ($D_{KL}[\cdot]$) is always non-negative. This KL divergence is between the probability density over external states η, given the MB state b, and a variational density $q_\mu(\eta)$ over external states parameterized by the internal state μ. Note here that the blanket states b are considered components of the "particle" state π; hence the VFE defined by Eq. (4.1) is the VFE "for" or "experienced by" S. We could, however, also consider a composite state $\rho = (\eta, b)$ and write an analog of Eq. (4.1) in which the roles of μ and η are exchanged; the resulting $F(\rho)$ would be the VFE "for" or "experienced by" E.

If the joint system $S \sqcup E$, where $\sqcup$ denotes disjoint union, is represented as a random dynamical system, then S having an MB requires that state trajectories that start in μ remain in μ; the states of S cannot, in other words, diverge to outside of the boundary of S. This condition is met if S has some *non-equilibrium steady state* (NESS) density. If we view the internal state μ as encoding a posterior probability distribution over the external state η, then minimizing VFE is, effectively, minimizing a prediction error under a generative model (GM) encoded by the NESS density. In this interpretation, Eq. (4.1) may be viewed as defining a "Bayesian mechanics" [44, 45] and minimization of VFE is a form of inference, termed *active inference* because one way of minimizing VFE is to act on the environment to move it toward an expected state. In this case, an agent S is competent if and only if it is an effective minimizer of its experienced VFE, i.e. if it can prevent its VFE from becoming high enough to drive its states far from its NESS, destroying the integrity of its MB [40].

Classical dynamical systems that are local when embedded in spacetime, i.e. compliant with Special Relativity, only transfer information casually; hence they can be represented as causal networks as shown in Fig. 4.5. This assumption of locality is, operationally, an assumption that the embedding spacetime is simply connected [46], as discussed in Chap. 7 below. Local states of a local, classical, dynamical system can, therefore, only depend on the "context" defined by more distant states via causal information transfer. Some systems, however, do not behave in this way. We cannot, therefore, understand information flow in generic systems using only the classical FEP.

Box 4.3: Physics and neuroscience

The FEP was originally formulated as a theory of the functional architecture of the mammalian brain [47–51]. Modeling the brain—or the organism—and its embedding environment as a dynamical system satisfying the Langevin and Fokker-Planck equations, interposing a boundary between the two that functions as an MB, and interpreting gradient descent on VFE as inference continue a tradition of applying physical principles to neuroscience that can be traced back to the pioneering works of Ramón y Cajal [52], Sherrington [53], McCulloch and Pitts [54], Ashby [55], and Hodgkin and Huxley [56]. Biophysics, from this historical perspective, is an outgrowth from and generalization of physics as applied to neuroscience. The "physics of complexity" lineage from Schrödinger [57] to Wiener [58], Haken [59], and Prigogine [60] blended with the lineage from neuroscience to create biophysics as we know it today.

As noted in Chap. 1, the idea that what brains are doing is computing dates back at least to McCulloch and Pitts [54]. In the spirit of [17], it can be fomulated as a set of simple, architecture-independent claims [61]:

(1) The nervous system is an I/O system.
(2) The nervous system is a *functionally organized* I/O system.
(3) The nervous system is a *feedback control*, functionally organized, I/O system.
(4) The nervous system is an *information-processing*, feedback control, functionally organized, I/O system.
(5) Therefore, the nervous system is a computing system.

That feedback control is inherently hierarchical is evident from Fig. 4.1. Even "flat" feedforward systems such as ANNs, however, experience effective feedback control from their environments, just in consequence of the symmetry of physical interactions.

The FEP adds to the above an interpretation—a model in the sense of Fig. 1.2—of VFE as uncertainty and of VFE minimization as Bayesian satisficing. As applied to neuroscience, the FEP and the related, less formalized predictive-processing framework make the following broad claims [40, 47–50, 62, 63]; see also [44, 64–76] and, in each case, references therein.

(1) Nervous systems generate and test predictions, and respond to prediction errors by acting on themselves—via learning—or on their environments, i.e. on the bodies that support them. Measurable cortical responses, whether bioelectric, biomolecular, or biomechanical, indicate implementation of these fundamental processes.

(2) Nervous system development and learning construct and maintain a generative model of the measured environment, including the supporting body. Neuronal and nervous system processes, from gene regulation and cellular homeostasis to synaptic development, pruning, and remodeling to network-scale dynamics and cycles such as sleep implement construction and maintenance of the GM.

(3) Processes at every scale implement the same dynamics, with transitions between scales representable as *Renormalization Group* (RG) operators. Hierarchical systems implement renormalizable GMs that can be viewed as discrete homologues of deep convolution networks.

(4) Neurons interact with their environments via their cell membranes, through which both all metabolic resources and all communicative signals, whether chemical, electrical, or mechanical, flow. Pre- and postsynaptic channels, pumps, and receptors are embedded in their membranes, and membrane potentials are waves of electrically-coupled biochemical activity that propagate along the membrane. These information flows can all be represented as dynamic encodings on the neuron's MB. Information flows between cortical columns and their environments, including other cortical columns, can similarly be represented as dynamic encodings on the MBs of the columns.

(5) Physics is particularly relevant to the study of *synergetics*, as we recall, an interdisciplinary theory of model processes and mechanisms of spontaneous pattern formation with respect to environmental influences. The latter is vital for self-organization, since that depends on a system (considered as open) which is influenced by external energy sources. Principles of dynamical systems apply to show that an abrupt nonlinear bifurcation from one global pattern to another, is manifestly a correspondence between basins of attractors.

(6) The FEP, in partnership with predictive coding and hierarchical Bayesian inference, provides a methodology for studying allostatsis, interoception, memory, embodiment, and synergetics. To these we can add such phenomena as perception, emotions and sentience. From a clinical perspective, disorders of inference/prediction have been hypothesized as possible root causes for a range of cognitive conditions.

(7) Experimental methods that access information about neuronal state *in vivo*, from direct biochemical probes such as optogenetics or bioelectric probes such as patch clamps to lower-resolution, multineuron probes such as neuroimaging, can all be represented as actions by the environment on the neuron to which the neuron responds by acting back on the environment.

It is now well-established that non-neuronal cells also engage in bioelectric signalling, at time scales roughly three orders of magnitude slower than neurons [77–81]. Computational analysis and simulation of cell-cell communication, including the behavior of large neural networks is a vast, ever-expanding area of computer science, involving a spectrum of techniques at different levels of abstraction and with goals ranging from realistic modeling of biological function to the development of practical ANN architectures for general machine-learning applications. The extent to which biological systems can be seen as implementing standard algorithms such as error back-propagation remains an open question; see e.g. [82] for a recent review.

4.5 Causal Versus Non-causal Context Dependence

John Bell [83] and Simon Kochen and Ernst Specker [84] independently proved that generic quantum systems can exhibit non-causal context dependence; see [85] for a comparative analysis of these results. To get the relevant intuitions, it is useful to go back to pre-Einsteinian classical physics. In the physics of Newton and Lagrange, information transfer via the gravitational field was instantaneous; no time was required for a change in the orbit of Jupiter, for example, to affect the orbit of Earth [86]. Hence in pre-Einsteinian classical physics, the Earth's orbit depended instantaneously on the physical "context" defined by the orbit of Jupiter. Einstein's introduction of a finite speed of information transfer—via a finite speed of light—made this dependence non-instantaneous, and hence local or "causal" in the current meaning of that term. The Bell and Kochen-Specker theorems show that the states of quantum systems can depend on contexts that are distant in either space or time in a way that is non-causal, i.e. not explicable by information-transfer processes that require finite time.

The simplest exemplar of non-causal context dependence is two-qubit quantum entanglement demonstrated via a Bell/EPR experiment [87, 88]. The experiment involves a source of entangled pairs, e.g. photons or electrons with entangled spins, that are emitted in opposite spatial directions at or near the speed of light, and then detected by observers at macroscopic distances from the source, as shown in Fig. 4.6. The observers, Alice and Bob, select one of two designated "settings" on their detectors, e.g. spin directions for Stern-Gerlach interferometers, and make binary-valued measurements in less time than would be required for them to communicate classically. The result of multiple measurement rounds is a collection of outcome values of four variables, conventionally called A_1, A_2, B_1, and B_2 and allowed values of ± 1. As originally proven by Bell [89], a statistical analysis of these values can determine whether the results indicate quantum entanglement or only classical correlation. The most common test is the CHSH expectation value given by Eq. (2.1), which we repeat here for convenience:

$$EXP = |\mathbf{e}(S_1, E_1) + \mathbf{e}(S_1, E_2) + \mathbf{e}(S_2, E_1) - \mathbf{e}(S_2, E_2)| \qquad (4.2)$$

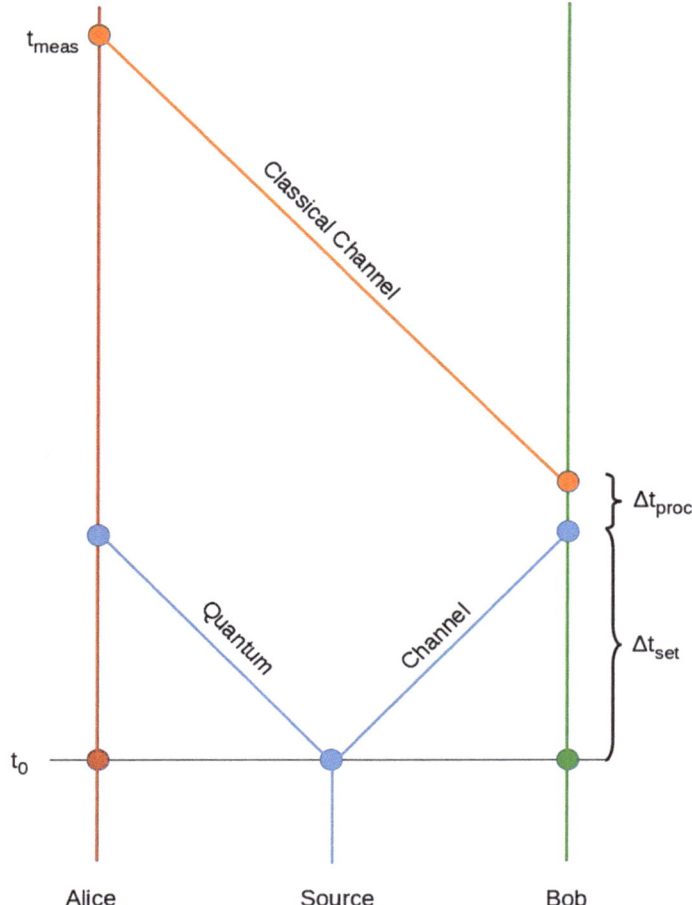

Fig. 4.6 A typical Bell/EPR experiment described in the lab frame. Sharing of measurement results via a classical channel is required to observe a Bell-inequality violation. If Alice's interaction with Bob's message is viewed as an ordinary quantum measurement, the entanglement is observable only by a third party. From [92] Fig. 7; CC-By license

where $\mathbf{e}(x, y)$ denotes the expectation value for a collection of joint measurements of x and y. If $EXP > 2$, classical data reported by A and B violate the CHSH inequality, indicating entanglement between q_A and q_B [90]. For single qubits, the upper limit is $EXP \leq 2\sqrt{2}$, which is the relevant Tsirelson bound [91].

Note that in any such multi-observer experiment, the results must be communicated between observers, or communicated to some third party, by classical means as discussed in Sect. 2.6; this classical channel is included in Fig. 4.6. We will see in the next chapter that this classical communication step is not just a practical component of non-causal context dependence, but a principled outcome of non-commutativity between Alice's and Bob's measurements.

References

1. Gould, J. L. Animal cognition. Current Biology 2004; 14(10): R372–R375.
2. Maturana, H. R. and Varela, F. J. Autopoiesis and cognition: The realization of the living. Reidel, Boston (1980).
3. Baluška, F. and Levin, M. On having no head: cognition throughout biological systems. Front Psychol 2016;7:902.
4. Levin, M. Technological approach to mind everywhere: An experimentally-grounded framework for understanding diverse bodies and minds. Front Syst Neurosci 2023;16:768201.
5. Baars, B.J. (1983). Conscious Contents Provide the Nervous System with Coherent, Global Information. In: Davidson, R.J., Schwartz, G.E., Shapiro, D. (eds) Consciousness and Self-Regulation. Springer, Boston, MA. https://doi.org/10.1007/978-1-4615-9317-1_2
6. Dehaene, S. and Naccache, L. Towards a cognitive neuroscience of consciousness: Basic evidence and a workspace framework. *Cognition* 79 2001, 1–37.
7. Micali, G. and Endres, R.G. Bacterial chemotaxis: Information processing, thermodynamics, and behavior. Curr. Opin. Microbiol. 2016, 30, 8-15.
8. Fields, C. and Levin, M. How do living systems create meaning? Philosophies 2020; 5: 36.
9. Kuchling, F., Fields, C. and Levin, M. Metacognition as a consequence of competing evolutionary time scales. Entropy 2022; 24: 601.
10. Erwin, D. H. and Davidson, E. H. The evolution of hierarchical gene regulatory networks. Nature Reviews Genetics 2009; 10: 141-148.
11. Hatleberg, W. L. and Hinman, V. F. Modularity and hierarchy in biological systems: Using gene regulatory networks to understand evolutionary change. Current Topics in Developmental Biology 2021; 141: 39-73.
12. Gibson JJ. The ecological approach to visual perception. Boston: Houghton-Miffin; 1979.
13. Michaels C, Carello C. Direct Perception. Englewood Cliffs, NJ: Prentice-Hall; 1981.
14. Brooks, R. A. Intelligence without representation. Artificial Intelligence, 1991; 47, 139-159.
15. Anderson, M. L. Embodied cognition: A field guide. Artificial Intelligence 2003; 149: 91-130.
16. Chemero A. Radical embodied cognitive science. Rev Gen Psychol 2013;17:145-150.
17. Horsman, C., Stepney, S., Wagner, R.C. and Kendon, V. When does a physical system compute? Proc. R. Soc. A 2014, 470, 20140182.
18. Tanaka, G., Yamane, T., Héroux, J. B., Nakane, R., Kanazawa, N., Takeda, S., Numata, H., Nakano, D. and Hirose, A. Recent advances in physical reservoir computing: A review. Neural Networks 2019; 115: 100-123.
19. Fields, C., Albarracin, M., Friston, K., Kiefer, A., Ramstead, M. J. D. and Safron, A. How do inner screens enable imaginative experience? Applying the free energy principle directly to the study of conscious experience. Neuroscience of Consciousness 2025; 2025: niaf009.
20. Baars, BJ (1997) In the theatre of consciousness. Global Workspace Theory, a rigorous scientific theory of consciousness. J. Cons. Studies 4(4), 292–309.
21. Baars, B. J. and Franklin, S. How conscious experience and working memory interact. *Trends in Cognitive Science* 7 2003, 166–172.
22. Baars BJ, Franklin S, Ramsoy TZ (2013) Global workspace dynamics: Cortical "binding and propagation" enables conscious contents. Front Psychol 4:200.

23. Dehaene S, Sergent C, Changeux J-P (2003) A neuronal network modeling subjective reports and objective physiological data during conscious perception. Proc Natl Acad Sci USA 100(14):8520–8525.
24. Dehaene S, Charles L, King J-R, Marti S (2014) Toward a computational theory of conscious processing. Curr Opin Neurobiol 25:76–84.
25. Hommel, B. (2004). Event files: Feature binding in and across perception and action. Trends in Cognitive Sciences, 8, 494–500.
26. Seth AK (2013) Interoceptive inference, emotion, and the embodied self. Trends Cognit Sci 17(11):565–573.
27. Fields, C. and Glazebrook, J.F. Representing measurement as a thermodynamic symmetry breaking. Symmetry 2020, 12, 810.
28. Wallace, R. *Consciousness: A Mathematical Treatment of the Global Neuronal Workspace.* Springer, New York, USA, 2005.
29. Shanahan, M. The brain's connective core and its role in animal cognition. *Phil. Trans. Royal Soc. B 367* **2012**, 2704–2714.
30. Shepherd, G. M. G. and Yamawaki M. (2021) Untangling the cortico-thalamo-cortical loop: cellular pieces of a knotty circuit puzzle. Nature Reviews Neuroscience 22, 389–406.
31. Whyte, C. J., Redinbaugh, M. J., Shine, J. M. and Saalmann, Y. B. (2024) Thalamic contributions to the state and contents of consciousness. Neuron 112, 1611–1625.
32. Franklin, S., Ramamurthy, U., DMello, S. K., McCauley, L., Negatu, A., Silva, R. and Datla, V. (2007) LIDA: A computational model of global workspace theory and developmental learning', AAAI Fall Symposium on AI and Consciousness: Theoretical Foundations and Current Approaches, pp. 61–66.
33. MacKay, D. J. C. (1992) A practical Bayesian framework for backpropagation networks. Neural Computation 4, 448–472.
34. Fields, C. and Glazebrook, J. F. (2019) A mosaic of Chu spaces and Channel Theory II: Applications to object identification and mereological complexity. Journal of Experimental and Theoretical Artificial Intelligence 31, 237–265.
35. Knill, D. C., Pouget, A. (2004) The Bayesian brain: The role of uncertainty in neural coding and computation. Trends Neurosci. 27, 712–719.
36. Sanborn, A. N., Chater, N. (2016) Bayesian brains without probabilities. Trends Cognit. Sci. 20, 883-893.
37. Huang, Y., Rao, R. P. N. (2011) Predictive coding. WIRES Cogn. Sci. 2, 580–593.
38. Spratling, M. W. (2017) A review of predictive coding algorithms. Brain Cogn. 112, 92-97.
39. van Lieshout, L. L. F., Zhang, Z., Friston, K. J., Bekkering, H. (2025) Predictive processing: Shedding light on the computational processes underlying motivated behavior. Behav. Brain Sci. 48, in press.
40. Friston, K.J. A free energy principle for a particular physics. arXiv 2019, arXiv:1906.10184.
41. Pearl, J. Probabilistic Reasoning in Intelligent Systems: Networks of Plausible Inference; Morgan Kaufmann: San Mateo, CA, USA, 1988.
42. Fields, C. and Levin, M. Multiscale memory and bioelectric error correction in the cytoplasm–cytoskeleton-membrane system. WIREs Syst Biol Med 2017, e1410.
43. Feynman, R.P. Statistical Mechanics; Benjamin: Reading, MA, USA, 1972.
44. Ramstead, M.J., Sakthivadivel, D.A.R., Heins, C., Koudahl, M., Millidge, B., Da Costa, L., Klein, B. and Friston, K.J. On Bayesian mechanics: A physics of and by beliefs. Interface Focus 2022, 13, 20220029.
45. Friston, K.J., Da Costa, L., Sakthivadivel, D.A.R., Heins, C., Pavliotis, G.A., Ramstead, M.J. and Parr, T. Path integrals, particular kinds, and strange things. Phys. Life Rev. 2023, 47, 35-62.
46. Fields, C., Glazebrook, J. F., Marcianò, A., and Zappala, E. (2025) ER = EPR is an operational theorem. Physics Letters B 860, 139150.

47. Friston, K. The free-energy principle: A unified brain theory? Nat. Rev. Neurosci. 2010, 11, 127-138.
48. Friston, K. J., Penny, W., Phillips, C., Kiebel, S., Hinton, G. and Ashburner, J. Classical and Bayesian inference in neuroimaging: Theory. *Neuroimaging* 16(2) 2002, 465–483.
49. Friston K.J. A theory of cortical responses. *Philos Trans R Soc Lond B, Biol Sci* 360 (2005), 815–36.
50. Friston K. J., Kilner, J., and Harrison, L. A free energy principle for the brain. J Physiol Paris 2006;100:70–87.
51. Friston K. J. and Stephan, K. E. Free-energy and the brain. *Synthese* 159 (2007), 417–58.
52. Newman, E. A., Arague, A. and Dubinsky, J. M. (eds) em The beautiful brain: the drawings of Santiago Ramón de Cajal. (contributors: Swanson, L. W., King. L. S. and Himmel, E.) Abrams, New York, 2017.
53. Eccles, J. C. and Gibson, W. C. *Sherrington: His LIfe and Thought*. Springer International, Berlin, Heidelberg, New York, 1979.
54. McCulloch, W. S., Pitts, W. (1943) A logical calculus of the ideas immanent in nervous activity. Bulletin of Mathematical Biophysics. 5(4), 115-133.
55. Ashby, W. R. *Design for a Brain*. Chapman and Hall, London, 1952,
56. Hodgkin, A. L. and Huxley, A. F. A quantitatve description of membrane current and its application to conductance and excitation in nerve. *The Journal of Physiology* 117(4) (1952), 500–544.
57. Schrödinger, E. *What is Life?* (1944). Reprinted in *What is Life? And Mind and Matter*. Cambridge Univ. Press, Cambridge UK, 2006.
58. Wiener, N. *Cybernetics: Or Control and Communication in the Animal and the Machine*. Hermann & Cie, Paris, 1948.
59. Haken, H. *The Science of Structure: Synergetics*. Van Nostrand Reinhold, New York, 1984.
60. Prigogine, I. *From Being to Becoming : Time and Complexity in the Physical Sciences*. Freeman, San Francisco, 1980.
61. Piccini, G. and Bahar, S. Neural computation and the computational theory of cognition. *Cognitive Science* 34 (2013), 453–488.
62. Friston, K. Life as we know it. J. R. Soc. Interface 2013, 10, 20130475.
63. Friston, K. FitzGerald, T., Rigoli, F., Schwartenbeck, P., O'Doherty. J. and Pezzulo, G. Active inference and learning. *Neuroscience and Biobehavioral Reviews* 68 (2016), 862–879.
64. Kirchoff, M., Parr, T., Palacios, E., Friston, K. and Kilverstein, J. The Markov blankets of life: autonomy, active inference and the free enrgy principle. *J. R. Soc. Interface* 15 20170792.
65. Hipólito, I., Ramstead, M. J. D., Convertino, L., Bhat, A., Friston, K. and Parr, T. Markov blankets in the brain. *Neurosci. Biobehv. Rev.* 125 (2021), 88–97.
66. Dayan, P., Hinton, G. E., Neal, R. M. and R. S. Zemel. The Helmholtz machine. *Neural Computation* 7 (1995), 1022–1037.
67. Barrett L. F., Tugade, M. M. and Engle, R. W. (2004). Individual differences in working memeory capacity and dual-process theories of the mind. *Psychol. Bull.* 130(4), 553–573.
68. Barrett, L. F. and Simmons, W. K. (2015) Interoceptive predictions in the brain. Nat. Rev. Neurosci. 16(7): 419–429.
69. Barrett, L. F. (2017). The theory of constructed emotion: an active inference account of interoception and categorization. *Social Cognitive and Affective Neuroscience* 12(1), 1–23.
70. Ramstead, M. J. D., Badcock, P. D. and Friston, K. J. Answering Schrödinger's question. A free-energy formulation. *Physics of Life Reviews* 24 (2018), 1–16.
71. Seth, A. K., Barrett, A. B. and Barnett, L. Causal density and integrated information as measures of conscious level. *Phil. Trans. R. Soc.* 369 2011, 3748–3767.
72. Seth, A. K., Suzuki, K. and Critchley, H. D. (2012). An interoceptive predictive coding model of conscious presence. *Frontiers in Psychology* 2 Article 395, 16 pp. https://doi.org/10.3389/fpsyg.2011.00395

73. Seth, A. K. 2014 The cybernetic Bayesian brain: from interoceptive inference to sensorimotor contingencies. In: T. Metzinger & J. M. Windt (Eds). *Open MIND*. Frankfurt am Main: MIND Group, Ch. 35.

74. Seth, A. K., Friston, K. J. 2016 Active interoceptive inference and the emotional brain. *Philos. Trans. R. Soc. Lond. B* 371(1708), 20160007.

75. Ciompi, L and Tschacher, W. Effect-logic, embodiment, synergetics, the free energy principle: new approaches to the understanding and treatment of schizophrenia. *Entropy* 23 (2021), 1619.

76. Von Toussaint, U. Bayesian inference in physics. *Rev. Mod. Phys.* 83 (2011), 943.

77. Levin, M. Molecular bioelectricity: how endogenous voltage potentials control cell behavior and instruct pattern regulation in vivo. *Mol. Biol. Cell.* 2014;25:3835–3850.

78. Pezzulo G, Levin M. Re-membering the body: Applications of computational neuroscience to the top-down control of regeneration of limbs and other complex organs. *Integr. Biol.* 2015;7:1487–1517.

79. Levin, M. The computational boundary of a "Self": Developmental bioelectricity drives multicellularity and scale-free cognition. *Front. Psychol.* 10 (2019), 2688.

80. Levin, M. Bioelectric signaling: Reprogrammable circuits underlying embryogenesis, regeneration, and cancer. *Cell* 2021;184:1971–89.

81. Levin, M. Darwin's agential materials: Evolutionary implications of multiscale competency in developmental biology. *Cell. Mol. Life Sci.* 80 (2023), 142.

82. Wright, L. G., Onodera, T., Stein, M. M. et al. Deep physical neural networks enabled by backpropagation algoritms for arbitrary physical systems. *Nature* 601 (2022), 549–555.

83. Bell, J. S. (1966). On the problem of hidden variables in quantum mechanics. Reviews of Modern Physics 38, 447–452.

84. Kochen, S., Specker, E. P. (1967). The problem of hidden variables in quantum mechanics. Journal of Mathematics and Mechanics 17, 59–87.

85. Mermin, N. D. (1993). Hidden variables and the two theorems of John Bell. Reviews of Modern Physics 65, 803–815.

86. Gisin, N. Can relativity be considered complete? From Newtonian nonlocality to quantum nonlocality and beyond. arxiv:quant-ph/0512168v1, 2005.

87. Aspect, A.; Graingier, P. and Roger, G. Experimental realization of Einstein-Podolsky-Rosen-Bohm *Gedankenexperiment*: A new violation of Bell's inequalities. Physical Review Letters 1982; 49(2): 91-94.

88. Aspect, A., Grangier, P. Roger, G.: Experimental tests of realistic local theories via Bell's theorem. Phys. Rev. Lett. 47, 460–463 (1981).

89. Bell, J. S. (1964). On the Einstein–Podolsky-Rosen paradox. Physics 1, 195–200.

90. Clauser, J.F., Horne, M.A., Shimony, A. and Holt, R.A. Proposed experiment to test local hidden-variable theories. Phys. Rev. Lett. 1969, 23: 880-884.

91. Cirel'son, B.S. Quantum generalizations of Bell's inequality. Lett. Math. Phys. 1980, 4, 93–100.

92. Fields, C., Glazebrook, J.F. and Marcianò, A. Reference frame induced symmetry breaking on holographic screens. Symmetry 2021, 13, 408.

Non-commuting CCCDs and Non-causal Context Dependence

5

Observations and actions do not take place in a vacuum: what is seen and done is always seen and done in some *context*. Intuitively, the context of an observation or action comprises what has *not been observed* about the environment—whether due to technical limitations or attentional focus—when the observations or actions of interest are carried out. "Context effects" are ubiquitous in science; isolated experimental environments, controls, and replication are all strategies to minimize them.

As discussed in Chap. 4, context effects can be divided into two categories. *Causal*, or classical, or extrinsic context effects follow from ordinary causal interactions between unobserved components of the environment E and the components of E that are "of interest" to some observer. Recall that interactions are "causal" if, and only if, they comply with Special Relativity, i.e. if and only if all information flows forward in time at less than or equal to the speed of light. *Non-causal*, or quantum, or intrinsic context effects violate this condition.

The theory of non-causal context effects—sometimes just written "contextuality"—was originally developed by explicitly constructing observational scenarios in which sets of pairwise-commuting measurements produced violations of "local realism" when performed together [1]; see [2, 3] for relevant early work and [4, 5] for more recent examples. Local realism is the joint assumption of causality and "counterfactual definiteness", i.e. that all degrees of freedom have some particular value at all times, whether it is measured or not; see [6] for discussion. It is now known, however, that non-causal context effects will appear in any measurement setting involving a collection of physical degrees of freedom—formally, random variables—that can have values over which no well-defined joint probability distribution exists. A number of formulations of the theory have been developed, including sheaf-theoretic [7–10] (the sheaf concept is defined in Appendix Sect. D.1), graph-theoretic [11–14], Contextuality by Default [15–17], and others [18, 19]; see [20] for a comparative

C. Fields and J. Glazebrook, *Distributed Information and Computation in Generic Quantum Systems*, Synthesis Lectures on Engineering, Science, and Technology, https://doi.org/10.1007/978-3-031-97263-8_5

philosophical discussion and [21] for a particularly clear recent review. Non-causal context effects are now recognized as a significant resource for quantum computation [22–27].

Box 5.1: Contextuality and "quantum cognition"

One of the major contenders for a general theory of non-causal context effects, the Contextuality by Default (CbD) approach, was developed by mathematical psychologists [15–17]. Studies of human language use and decision making often reveal violations of standard probabilistic reasoning, many of which have been interpreted as evidence of quantum—or "quantum-like"—effects in human cognition [28–30]. Dzhafarov, Zhang and Kujala [31] showed in 2016 that the results of many such studies could, in fact, be explained in terms of causal context dependence. The CbD formalism is based on the assumption that *all* measurements potentially involve context dependence—an excellent assumption not just in psychology, but throughout the biosciences—and provides formal criteria for distinguishing causal from non-causal context dependence (see [32] for how meanings can be distorted by linguistic ambiguities and discrepancies; here the techniques of CbD and the sheaf-theoretic methods are partially combined for analyzing word and sentence meaning in some context). As shown in [21], causal context dependence can be represented as an additive correction to the CHSH inequality discussed in Sect. 4.5 that makes detection of quantum effects more difficult. Two experiments designed using CbD have unambiguously demonstrated non-causal context effects in human decision making [33, 34]. The distinction between causal and non-causal context dependence remains, however, controversial within psychology and has little effect on experimental work; see [35] for a recent review.

5.1   CCCDs and Contextuality

Here we adopt an observer-centered approach to non-causal context dependence motivated by idea that quantum theory is a theory of measurement as discussed in Box 2.1. This has the advantage of providing a complete account of non-causal context dependence with minimal ontological assumptions. We show, in particular, that non-causal context effects arise in any measurement setting that involves QRFs, represented as CCCDs, that do not mutually commute. If non-commutative QRFs are applied to a data set, in other words, a well-defined joint outcome probability distribution does not exist over that data set.

5.1.1   Observables in Context

To commence, we recall some details from [36] for building up the necessary classifiers. Consider the following (in practice, finite) sets:

(i) A is a set of "events" of observed value combinations, as related to

(ii) B a set of conditions specifying "objects/contents", or "influences", and

(iii) R a set of "contexts", or a set of "detectors", or "methods".

The set B can be decomposed $B = B^M \cup B^C$ (disjoint union), where B^M contains "objects/contents", or "degrees of freedom" that are observed, or measured in some event $a \in A$, and B^C contains what is *not* observed in the events in A. A priori assuming all events to be context-independent, we take the restriction $A|B^M$, implying that the observed events depend only on the observed conditions. Effectively, this assumes having local, complete measurements as that prevails in classical physics, or (equivalently) assuming a task-environment circumscription as a familiar concept in AI [37]. Thus, we consider the expanded space:

$$X := B \times R = (B^M \cup B^C) \times R \tag{5.1}$$

from which an appropriate classifier can be constructed.[1]

Definition 5.1.1 *An observable in context* is a classifier $\mathcal{A}$ of the form

$$\mathcal{A} = \langle A, X, \Vdash_A \rangle \tag{5.2}$$

Postulating a local logic $\mathcal{L}_\alpha = \mathcal{L}(\mathcal{A}_\alpha)$ for each classifier $\mathcal{A}_\alpha$, leads to a flow of logic infomorphisms:

$$\cdots \leftrightarrows \mathcal{L}_{\alpha+1} \leftrightarrows \mathcal{L}_\alpha \leftrightarrows \mathcal{L}_{\alpha-1} \leftrightarrows \cdots \tag{5.3}$$

providing semantic consistency within the information flow of a typical CCCD as depicted in (3.15).

When a CCCD commutes, each component classifier is said to represent a *co-deployable* observable, these being the observables encoding (commuting) QRFs in specifying what states are observed, and how they are observed. Otherwise, when a CCCD fails to commute, the variables in question are considered as *non-co-deployable*. Let us first consider the commutative case. Let $\mathcal{A}_1, \mathcal{A}_2, \ldots \mathcal{B}$ and $\mathcal{B}, \mathcal{C}_1, \ldots \mathcal{C}_k$, be finite sets of co-deployable observables, and let $\mathbf{C_1}$ and $\mathbf{C_2}$ be the respective cocone cores of the contexts they define. The combined set $\mathcal{A}_1, \mathcal{A}_2, \ldots \mathcal{B}, \mathcal{C}_1, \ldots \mathcal{C}_k$, of observables is co-deployable, in the sense of having a well-defined joint probability distribution, if and only if a core $\mathbf{C}$, and maps ϕ and ψ exist such that the corresponding CCCD commutes:

---

[1] The set $X := B \times R = $ (conditions, contexts) is a set basic to CbD [15, 16], but unlike that latter theory, we do not assume connectivity conditions, or particular systems of random variables. We are dealing here with a different approach altogether as based on distributed information flow and logic.

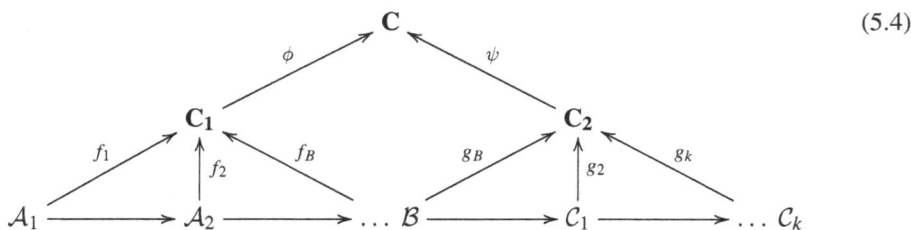

(5.4)

The failure of (5.4) to commute is a failure at the operator level, and hence the observables are considered to be *non-co-deployable*. The upshot of this is established in [36, Theorem 7.1].

Theorem 5.1.1 *With respect to a diagram of the form* (5.4), *failure of commutativity is equivalent to the non-existence of a consistent probability distribution across the combined set of observables* $\{\mathcal{A}_i, \mathcal{B}, \mathcal{C}_j\}$.

The containment of non-co-deployable observables within a diagram of the form of Diagram (5.4) that does not commute, and therefore is not a CCCD, amounts to *intrinsic contextuality* as prescribed by Theorem 5.1.1. Non-commuting CCCDs correspond to non-commuting QRFs that exhibit quantum contextuality when deployed in alternation. Examples in relationship to calibration and measurement, and to system identification and settings, are outlined in [36, Sect. 7] (see also [38]).

5.1.2   The Frame Problem and Intrinsic Contextuality

We next recall the Frame Problem (FP) studied within the distributed information flow of CCCDs in terms of an *informationally unencapsulated process* (IUP) as it features in cognitive science and AI [39]. An IUP is a means of assimilating all of the information from any source which is relevant *in fact* for completing a task, or for solving a problem. A priori, such a process is unlimited by circumscriptions of relevance, and hence requires, in principle, an unlimited availability of resources. Since an IUP is assumed to be programmed to know all relevant consequences of some action, it in principle solves the FP. The begging question is whether such processes exist?

> **Box 5.2: The Frame Problem**
>
> The FP was first stated explicitly by McCarthy and Hayes in 1969 [40]: given explicit
> representations of a set of facts—i.e. a "world"—a set of beliefs held by some agent,
> and an action by that agent, is there a way to efficiently represent the subset of beliefs
> that the agent need not update as a result of the action? Does, for example, the size
> of my desk change when I hit the "t" key on my laptop? What about the size of my
> office, or the color of the walls, or the positions of the pictures? Listing the facts that
> do change requires a complete causal model, while listing the facts that do not change
> appears to require "common sense"—whether I type a "t" is causally irrelevant to
> virtually everything about my office, and no causal story about typing ever considers
> such things as the color of the walls. The difficulty—if not impossibility—of capturing
> this kind of common sense about causal effects lies at the heart of the FP [39].
>
> While the FP is tractable and often simply ignored in finitely-circumscribed or
> "encapsulated" task environments, it is intractable in unbounded or "open" task envi-
> ronments [37, 41]; indeed it reduces to the Halting Problem and is, therefore, unde-
> cidable in such environments [42]. The FP is, therefore, closely related to context
> dependence; it is unsolvable whenever the context of an action cannot be fully spec-
> ified, and is intractable when the context is specifiable but not efficiently specifiable.
> This remains true even if all context dependence is causal, as it is by definition in the
> classical settings in which the FP was originally defined.
>
> The FP can be given a quantum formulation as the problem of determining whether
> an action changes the entanglement entropy of the environment. This problem is unde-
> cidable [38], as is obvious from the "no-go" results rehearsed in Chap. 2, Sect. 2.7.

To investigate this in terms of an appropriate CCCD as below, we can ask if further infomorphisms ϕ and ψ can be found to complete to a (colimit) cocone as in (5.5). Putting it
another way, when does the diagram commute?

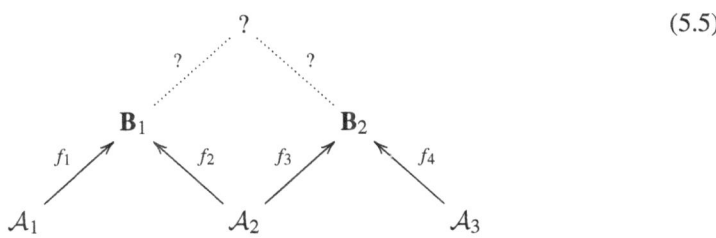

$$(5.5)$$

As the colimit classifier **C** that completes the diagram encompasses all of the information
from the component classifiers $\mathcal{A}_i$, the completed diagram represents an IUP, if and only
if the $\mathcal{A}_i$ gather all of the information that is *in fact* relevant to the problem (or task) in
question. The conditions to be satisfied are derived from [36, Theorem 7.1] which states that
the colimit exists, if and only if all $\mathcal{A}_i$ are co-deployable. This leads to [36, Corollary 8.1]: *a
distributed information-flow system such as in (5.5) represents an IUP only in the absence*

of intrinsic contextuality. That is, only if the diagram in (5.5) commutes. Failing that, there will be overall logical inconsistency in regulating the span of the problem, with the element of contextuality indicating that it is undecidable and/or intractable.

Regarding a CCCD as a graph (or network), and following Appendix D.1, a sheaf $\mathcal{F}$ can be defined over it, whereby the non-existence of a non-trivial global section for $\mathcal{F}$ implies noncommutativity [43]. This observation connects the behavior of a noncommutative CCCD with the approach of [7] showing that contextuality arises from the non-existence of a non-trivial global section for a given (conditional) probability distribution. More generally, the category $\mathfrak{C}$ of CCCDs [44] can be taken to be a *site* over which a sheaf $\mathcal{F}$ is definable, thus leading to the topos concept (see Appendix D.2), thus opening the door for further investigations to be pursued at a later date.

5.2   Context Dependence, the FP and the FEP

As described in the previous chapter, the FEP characterizes physical systems as prediction machines that, beyond trivial cases, implement GMs that model the actions of their environments on their boundaries. In the classical FEP, such GMs are restricted to causal models; indeed the classical FP characterizes the joint system-environment dynamics as a causal flow. All context dependence in the classical FEP is, therefore, causal. It is, moreover, in all but the simplest environments inevitable; no agent can measure the size or internal dynamics of its environment [45], so any assumptions of boundedness of causal effects or relevance built into its GM are at best heuristic.

An agent's GM is, or incorporates, its solution to the FP; the heuristic nature of GMs renders these FP solutions heuristic. This is evident in human problem solving, which incorporates what Drew McDermott called the "sleeping dog" heuristic: if X is not an obvious consequence of Y, assume that X is not a consequence of Y [46]. The consequences of using this appear often in the news; global climate change, for example, is still not universally recognized as a consequence of the Industrial Revolution. Restricted as it is to causal models, the classical FEP cannot represent non-causal context dependence. Observations of non-causal context dependence—e.g. in human decision making as discussed in Box 5.1—therefore motivate a quantum formulation of the FEP. This is the topic of the next chapter.

References

1. Kochen, S., Specker, E. P. (1967). The problem of hidden variables in quantum mechanics. Journal of Mathematics and Mechanics 17, 59–87.
2. Fine, A.: Hidden variables, joint probability and the Bell inequalities. Phys. Rev. Lett. 48, 291–295 (1982).

3. Peres, A.: Two simple proofs of the Kochen-Specker theorem, J. Phys. A: Math. Gen. 24, L175 (1991).
4. Cabello, A., Estebaranz, J. M., Garcia-Alcaine, G. Bell-Kochen-Specker theorem: A proof with 18 vectors. Phys. Lett. A 212 (1996) 183–187.
5. Spekkens, R. W.: Contextuality for preparations, transformations, and unsharp measurements. Phys. Rev. A 71, 052108 (2005).
6. Mermin, N. D. (1993). Hidden variables and the two theorems of John Bell. Reviews of Modern Physics 65, 803–815.
7. Abramsky, S., Brandenburger, A.: The sheaf-theoretic structure of non-locality and contextuality. New J. Phys. 13, 113036 (2011).
8. Abramsky, S., Brandenburger, A.: An operational interpretation of negative probabilities and no-signaling models. In van Bruegel, F. et al., eds, Panangaden Festschrift (Lecture Notes in Computer Science 8464) Springer, Switzerland, pp. 59–74 (2014).
9. Abramsky, S., Barbosa, R. S., Mansfield, S.: Contextual fraction as a measure of contextuality. Phys. Rev. Lett. 119, 050504 (2017).
10. Abramsky, S.: Contextuality: At the Borders of Paradox. In E. Landry, ed., Categories for the Working Philosopher, Oxford (2017) online edn, Oxford Academic, https://doi.org/10.1093/oso/9780198748991.003.0011
11. A. Cabello, S. Severini, A. Winter: Graph-Theoretic Approach to Quantum Correlations. Phys. Rev. Lett. 112, 040401 (2014).
12. Cabello, A., Kleinmann, M., Budroni, C.: Necessary and Sufficient Condition for Quantum State-Independent Contextuality, Phys. Rev. Lett. 114, 250402 (2015).
13. Cabello, A.: Simple method for experimentally testing any form of quantum contextuality, Phys. Rev. A 93, 032102 (2016).
14. Cabello, A.: Converting Contextuality into Nonlocality. Phys. Rev. Lett. 127, 070401 (2021).
15. Dzhafarov, E. N. and Kujala, J. V. (2017a). Contextuality-by-Default 2.0: Systems with binary random variables. In: J. A. Barros, B. Coecke and E. Pothos (eds.) *Lecture Notes in Computer Science* 10106, Springer, Berlin, 16–32.
16. Dzhafarov, E. N. Cervantes, V. H. and Kujala, J. V. (2017b). Contextuality in canonical systems of random variables.*Philosophical Transactions of The Royal Society* A 375, 20160389
17. Dzharfarov, E. N., Kon, M.: On universality of classical probability with contextually labeled random varaibles. J. Math. Psychol. 85, 17–24 (2018).
18. Gudder, S. Contexts in quantum measurement theory. Found. Phys. 49, 647–662 (2019).
19. Popescu, S.: Non-locality beyond quantum mechanics. Nature Phys. 10, 264–270 (2014).
20. Adlam, E. 2021 Contextuality, fine-tuning and teleological explanation. *Found. Phys.* 51 2021, 106 https://doi.org/10.1007/s10701-021-00516-y
21. Khrennikov, A. Contextuality, complementarity, signaling, and Bell tests. Entropy 2022, 24, 1380.
22. Raussendorf, R. Contextuality in measurement-based quantum computation. Phys. Rev. A 88, 022322 (2013).
23. Howard, M., Wallman, J., Veitch, V., Emerson, J.: Contextuality supplies the 'magic' for quantum computation. Nature 510, 351–355 (2014).
24. Bermejo-Vega, J., Delfosse, N., Browne, D. E., Okay, C., Raussendorf, R.: Contextuality as a resource for models of quantum computation with qubits. Phys. Rev. Lett. 119, 120505 (2017).
25. Frembs, M., Roberts, S., Bartlett, S. D.: Contextuality as a resource for measurement-based quantum computation beyond qubits. New J. Phys. 20, 103011 (2018).
26. Mansfield, S., Kashefi, E.: Quantum advantage from sequential-transformation contextuality. Phys. Rev. Lett. 121, 230401 (2018).
27. Amaral, B.: Resource theory of contextuality. Phil. Trans. R. Soc. A 377, 2019.0010 (2019).

28. Aerts, D. Quantum structure in cognition. J. Math. Psychol. 53 (2009) 314–348.
29. Pothos, E. M., Busemeyer, J. R. Can quantum probability provide a new direction for cognitive modeling? Behav. Brain Sci. 36 (2013) 255–327.
30. Khrennikov A (2015) Quantum-like modeling of cognition. Front. Phys. 3:77. https://doi.org/10.3389/fphy.2015.00077
31. Dzhafarov, E. N., Zhang, R., Kujala, J. (2016b). Is there contextuality in behavioural and social systems? Philosophical Transactions of The Royal Society A 374, 20150099.
32. Wang, D., Sadrzadeh, Abramsky, S. and Cervantes, V. H. On the quantum-like contextuality of ambiguous phrases. *Proceedings of the 2021 Workshop on Semantic Spaces at the Intersection of NLP, Physics and Cognitive Science*, pp. 42–52. Association for Computational Linguistics 2021.
33. Cervantes, V. H., Dzhafarov, E. N. (2018). Snow Queen is evil and beautiful: Experimental evidence for probabilistic contextuality in human choices. Decision 5, 193–204.
34. Basieva, I., Cervantes, V. H., Dzhafarov, E. N., Khrennikov, A. (2019). True contextuality beats direct influences in human decision making. Journal of Experimental Psychology: General 148(11), 1925–1937.
35. Bruza, P. D., Fell, L., Hoyte, P., Dehdashi, S., Obeid, A., Gibson, A., Moreira, C. Contextuality and context-sensitivity in probabilistic models of cognition. Cogn. Psychol. 140 (2023) 101529.
36. Fields, C.; Glazebrook, J. F. Information flow in context-dependent hierarchical Bayesian inference. *J. Expt. Theor. Artif. Intell.* 34 2022, 111–142.
37. Dietrich, E., Fields, C.: The role of the frame problem in Fodor's modularity thesis: A case study of rationalist cognitive science. In: K. M. Ford and Z. W. Pylyshyn, Eds., The Robot's Dilemma Revisited. Norwood, NJ: Ablex, pp. 9–24 (1996).
38. Fields, C.; Glazebrook, J. F. Separability, contextuality, and the quantum Frame Problem. International Journal of Theoretical Physics 2023; 62: 159.
39. Shanahan, M. The Frame Problem. Stanford Encyclopedia of Philosophy, 2016 (https://plato.stanford.edu/entries/frame-problem/; accessed 15 Feb 2025).
40. McCarthy, J., Hayes, P. J.: Some philosophical problems from the standpoint of artificial intelligence. In: Michie, D. and Meltzer, B., dds., Machine Intelligence, Vol. 4. Edinburgh: Edinburgh University Press, pp. 463–502 (1969).
41. Fodor, J. Modules, Frames, Fridgeons, Sleeping Dogs, and the Music of the Spheres. In Pylyshyn, Z. W. (Ed) The Robot's Dilemma: The Frame Problem in Artificial Intelligence, Norwood, NJ: Ablex, 1987.
42. Dietrich, E., Fields, C.: Equivalence of the Frame and Halting problems. Algorithms 13, 175 (2020).
43. Hansen, J. A gentle introduction to sheaves on graphs, 2021. https://jakobhansen.org/publications/gentleintroduction.pdf
44. Fields, C. Glazebrook, J. F.; Marcianò, A. The physical meaning of the Holographic Principle. Quanta 2022; 11:72–96.
45. Moore, E.F. Gedankenexperiments on sequential machines. In Autonoma Studies; Shannon, C.W., McCarthy, J., Eds.; Princeton University Press: Princeton, NJ, USA, 1956; pp. 129–155.
46. McDermott, D. We've Been Framed: Or Why AI Is Innocent of the Frame Problem. In Pylyshyn, Z. W. (Ed) The Robot's Dilemma: The Frame Problem in Artificial Intelligence, Norwood, NJ: Ablex, 1987.

The Quantum FEP

A quantum-theoretic formulation of the FEP (qFEP) was introduced in [1]; see [2] for a comparison of the classical and quantum formulations and [3–6] for applications of qFEP to generic systems. The fundamental distinction between the classical FEP and qFEP is in the representation of information flows between systems; these are continuous in the classical FEP and discrete in qFEP. Hence the classical FEP can be seen as the classical "macroscopic" limit of qFEP, the limit in which information flows are large enough to be considered continuous.

The basic formal moves from the classical FEP to qFEP are:

- The classical dynamical system assumed by the FEP is replaced by the internal Hamiltonian H_U of some isolated system U.
- The MB separating the system of interest S from its environment E is replaced by a holographic screen $\mathscr{B}$.
- The criterion of conditional statistical independence between S and E is replaced by the criterion of separability, i.e. that $|U\rangle = |S\rangle|E\rangle$ in Dirac's notation.

These formal changes alter nothing about the basic intuitions underlying the FEP, and preserve its status as a translation from the language of physics to the language of Bayesian satisficing. The qFEP remains a theory of interactions between mutually-distinguishable—i.e. separable—systems, and characterizes all such systems, as does the classical FEP, as behaving so as to preserve the integrity of their shared boundary $\mathscr{B}$. The qFEP expands the scope of the FEP, however, to include non-classical entities, e.g. quantum fields or Black Holes. Via the idea that generic quantum interactions can be viewed as "measurements" discussed in Box 2.1, the quantum formulation renders the FEP an even more general theory of information exchange. Indeed, it reveals the FEP as the general representation of agents

© The Author(s), under exclusive license to Springer Nature Switzerland AG 2026   87
C. Fields and J. Glazebrook, *Distributed Information and Computation in Generic Quantum Systems*, Synthesis Lectures on Engineering, Science, and Technology,
https://doi.org/10.1007/978-3-031-97263-8_6

interacting with their worlds that physicists have often sought under the rubric of a "theory of measurement".

6.1 Formal Statement of the qFEP

6.1.1 Action of the Hamiltonian H_{SE}

Adopting the notation of Chap. 2, let U be an isolated system that is factored as $U = SE$ where $|SE\rangle = |S\rangle|E\rangle$ over the time period of interest. The overall Hamiltonian is $H_U = H_S + H_E + H_{SE}$, where Eq. (2.2) specifies H_{SE} as:

$$H_{SE} = \beta_k k_B T_k \sum_{i=1}^{N} M_i^k \tag{6.1}$$

for $k = S$ or E, where k_B is Boltzmann's constant, T_k is temperature, $\beta_k \geq \ln 2$, and the operators M_i^k are Hermitian with eigenvalues ± 1. The interaction H_{SE} is defined at the boundary $\mathscr{B}$ between S and E, and can be interpreted as the alternate "writing" and "reading" of N-bit strings on and from $\mathscr{B}$, respectively, by both S and E as illustrated in Fig. 2.2. As discussed in Box 2.2, each of the M_i^k requires a local reference frame that defines the effective z-axis direction corresponding to a $+1$ outcome value. Recall from Eq. 2.2.1 that we can write:

$$M_i^k = \exp(\iota \phi_i^k(t_U)) \sigma_{\phi_i^k} \tag{6.2}$$

where $\phi_i^k(t_U)$ is the reference direction for M_i^k as a function of the background time t_U, and where $\sigma_{\phi_i^k}$ is the spin operator for the ϕ_i^k direction. Hence for constant N—i.e. effective boundary Hilbert-space dimension $\dim(\mathcal{H}_{\mathscr{B}}) = 2^N$—we can treat any time dependence of H_{SE} as given by the time dependence of the local QRFs ϕ_i^k. We therefore assume, as an explicit condition for the qFEP, that the size of the $S - E$ boundary, and hence N, changes only slowly relative to the characteristic time of H_{SE}. This is, effectively, an assumption that the overall dynamics H_U is slowly-varying, which enables the separability condition $|U\rangle = |S\rangle|E\rangle$ to remain satisfied over "macroscopic" times. This is clearly required if S and E are to be regarded as "persistent systems" as this term is defined within the classical FEP.

We know from Sect. 2.4 that the internal dynamics of both S and E can be described by specifying a collection of QRFs, and from Sect. 2.5.1 that non-commutative QRFs induce compartmentalization, and hence internal boundaries across which classical information is exchanged, in any system that implements them. Hence agents of interest will, in general, have the architecture shown in Fig. 4.2, with each of the PALs, the TFE distribution system, and the metaprocessor (MP) all implemented by compartmentalized QRFs. As described in Chap. 4, the task of the MP in this architecture is to regulate the lower-level QRFs, turning them "on" or "off"—i.e. deploying them—selectively in response to its measured

VFE. Deploying a QRF resets the values of the ϕ_i^k to those specified by that QRF; hence the time dependence of H_{SE} can be captured in the computational language of MP-driven, VFE-dependent, selective QRF deployment.

As VFE is defined at a single boundary, it suffices for defining VFE to consider agents that implement only mutually-commutative QRFs, which can be represented as commutative CCCDs and therefore combined into a single, compartmentalized QRF. In such "trivial" [1] or "atomic" [7] agents, data on both thermodynamic and informative sectors of the boundary are processed by the single available QRF. The informative and thermodynamic sectors of the exterior boundary of the agent as a whole—$\mathscr{B}_S$ in Fig. 2.2—are then emergent properties of the overall compartmentalized architecture. The behaviorally-relevant VFE for the whole system S is, in this architecture, the VFE measured by the MP at its boundary $\mathscr{B}_{MP}$.

6.1.2 Probabilities and Prediction Errors

While the classical FEP treats physical systems as Bayesian agents, the definition of VFE given by Eq. (4.1) employs "objective" probabilities, i.e. probabilities from the perspective of a theorist who can "observe" all of U. Using the notation of [8, Eq. 8.4], we can re-write the last of the definitions of the VFE $F(\pi)$ given by Eq. (4.1) as:

$$F(\pi) \triangleq \underbrace{\mathfrak{I}(\pi)}_{Surprisal} + \underbrace{D_{KL}[Q_\mu(\eta) \parallel P(\eta|b)]}_{Divergence} \geq \mathfrak{I}(\pi) \tag{6.3}$$

where as in Eq. (4.1) π is a "particular" state $\pi = (b, \mu)$ comprising MB (b) and internal (μ) components and η is the "external" or environmental state. The functional $F(\pi)$ is an upper bound on surprisal $\mathfrak{I}(\pi) = -\log P(\pi)$ because the Kullback-Leibler divergence (D_{KL}) term is always non-negative. This KL divergence is between the density over external states η, given the MB state b, and a variational density $Q_\mu(\eta)$ over external states parameterised by the internal state μ.

Following the development in [1], we can write the surprisal for a quantum system S in its most general form as:

$$\mathfrak{I}^S(t_U) = -P(|\mathscr{B}(t)\rangle \mid |S(t_U)\rangle) \tag{6.4}$$

and the corresponding evidence bound as:

$$D_{KL}[Q_{|\mathscr{B}(t_U)\rangle}(|E(t_U)\rangle) \parallel P(|E(t_U)\rangle \mid |\mathscr{B}(t_U)\rangle)] \tag{6.5}$$

In the current setting, however, these expressions have little direct utility, as our effective starting point, Eq. (6.1), constrains neither $|S(t_U)\rangle$ nor $|E(t_U)\rangle$, and neither is an observable for any system in the current setting. Characterizing prediction error *as measured by a system of interest*, therefore, requires abandoning objective probabilities.

Let S be an atomic agent, with a single QRF Q that processes data encoded on its boundary $\mathscr{B}$; in the simplest case, we can regard Q as simply returning a single logical bit valued as ± 1. A generative model for E as observed by S is, therefore, just a prior probability distribution over this binary variable. As discussed in [1], we can represent this prior probability distribution as a Markov kernel $\mathbb{M}_Q$, and represent its updating over discrete time steps k as the action of a "learning" operator:

$$\mathscr{L} : (\mathbb{M}_Q(k), \ [Q(k)]) \mapsto \mathbb{M}_Q(k+1) \tag{6.6}$$

where the notation $Q(k)$ denotes the outcome value returned by Q at time step k.

As the records $Q(k)$ are just bit strings and Bayesian coherence is guaranteed by the commutativity of the operators composing the QRF Q, the sequence $[Q(1)], [Q(2)], \ldots [Q(k)]$ can be represented by a "true" Markov kernel $\mathbb{M}(k)$ satisfying the following commutativity constraint:

$$
\begin{array}{ccc}
Q(i) & \xrightarrow{\ \mathbb{M}_E\ } & Q(i+1) \\[2pt]
{\scriptstyle Q}\big\uparrow & & \big\uparrow{\scriptstyle Q} \\[2pt]
|\mathscr{B}(t)\rangle & \xrightarrow{\ \mathcal{P}_U\ } & |\mathscr{B}(t+\Delta t_U)\rangle
\end{array}
\tag{6.7}
$$

where as in Fig. 2.4 one "tick" of S's internal clock corresponds to Δt_U externally and the notation $|\mathscr{B}\rangle$ indicates the state of the qubit array encoded on $\mathscr{B}$. At step k in S's acquisition of state information from $\mathscr{B}$, S's prediction error Er_Q can be represented as the difference between "true" and learned prior probability distributions:

$$Er_Q(k) = d(\mathbb{M}_Q(k), \mathbb{M}(k)) \tag{6.8}$$

where the d is the metric distance on Markov kernels. This definition is independent of $\mathscr{L}$ and hence of the sources of S's prediction errors.

6.1.3 Defining VFE

The error $Er_Q(k)$ given by Eq. (6.8) represents the *reducible* error attributable to Q. It is an upper bound on surprisal analogous, in the current setting, to $F(\pi)$. The error $Er_Q(k)$, therefore, represents S's reducible VFE. This error must be computed for each non-commuting QRF, and hence for each bounded compartment of a composite agent, as the Markov kernels appearing in Eq. (6.8) become undefined if collections of non-commuting operators are considered. We can, therefore, formulate the FEP for generic quantum systems as:

FEP: *A generic quantum system S will act so as to minimize Er_Q for each deployable QRF Q.*

Active inference—learning combined with action that modifies the environment—is the generic mechanism for reducing prediction error, i.e. VFE, and hence the generic mechanism by which systems satisfy the FEP. In composite systems, VFE is reduced in the short term via control, or in Bayesian terms, policy selection: choosing which QRFs to deploy, and hence both "what to look for" and "what to act on" under current circumstances.

Box 6.1: Tensor networks as generic representations of control

A tensor network (TN) is a factorization of a tensor, i.e. a multilinear map or high-dimensional matrix, into a network of tensors, in which the links in the network are tensor contractions.[a] TNs have recently found wide application to quantum many-body systems [10]. We showed in [3] that they provide a generic representation of control flow for systems compliant with the qFEP. In particular, we have:

Theorem 6.1.1 *A system S exhibits non-trivial control flow if, and only if, its control flow can be represented by a TN.*

Where "non-trivial" means that S can deploy multiple QRFs.

Suppose S deploys multiple, distinct QRFs $Q_1, Q_2, \ldots Q_n$, acting on its environment E, where $n \ll N = \dim(H_{SE})$. Classical control flow in S can then be represented by a matrix $\mathbf{CF} = [P_{ij}]$, where P_{ij} is the probability of the control transition $Q_i \rightarrow Q_j$. The matrix $\mathbf{CF}$ is a 2-tensor. Theorem 6.1.1 states that this tensor can be decomposed into a TN.

Proof *(Proof (Theorem 6.1.1))* Suppose first that control flow in a system S can be represented by a TN. A TN is, by definition, a factorization of a tensor operator into a network of tensor operators. This network can be either hierarchical or flat; if it is hierarchical, each layer can be considered a flat TN. Hence no generality is lost in considering just the case of a flat TN, which is an operator contraction $T = \ldots T_{ij} T_{jk} T_{kl} \ldots$, where summation on shared indices is left implicit. In general, $T_{jk} \neq T^T_{jk} = T_{kj}$, hence these expressions do not commute. They therefore represent non-trivial control flow. Conversely, any non-trivial control flow can be written, at any fixed scale or level of abstraction, as a linear sequence of (in general probabilistic) operators. The fixed order of operators in the sequence can be encoded formally by adding "spatial" indices as needed to allow contraction over shared indices. Hence any non-trivial control flow at a fixed scale can be written as a flat TN. This construction can be repeated at each larger scale to produce a hierarchical TN over a collection of "lowest-scale" TNs. □

We can now examine two corollaries of this result:

Corollary 6.1.1 *Decoherent reference sectors exist on a boundary $\mathscr{B}$ if and only if control flow can be implemented by a TN.*

Proof Decoherence between sectors requires independently-deployable, non-commuting QRFs. This requires a control structure that factors, hence by Theorem 6.1.1, it requires a TN. Conversely, a TN factors the control structure, making QRFs independently deployable, which renders their sectors decoherent.   □

Equivalently, the generative model implemented by a system factors if, and only if, control flow can be implemented by a TN.

Corollary 6.1.2 *The TN of any system compliant with the FEP is a decomposition of the Identity.*

Proof Classically, the FEP applies to systems with a NESS, a non-equilibrium steady state (NESS), and drives such systems to return to (the vicinity of) the NESS after any perturbation. Hence at a sufficiently large scale, the TN of any such system is a cycle, i.e., a decomposition of the Identity.   □

Theorem 6.1.1 provides a formal connection between the qFEP and generic formalisms for machine learning as discussed in [3]. We will return to some of these connections in Chap. 8 below.

---

[a] Formally, given a vector space V and some ground field $\mathbb{K}$, a *tensor of type (p,q)* is a multilinear map

$$T : \underbrace{V^* \times \cdots \times V^*}_{p \text{ copies}} \times \underbrace{V \times \cdots \times V}_{q \text{ copies}} \longrightarrow \mathbb{K}$$

defining T as a rank $(p+q)$-tensor. We can also specify T as an element of a *tensor product space*:

$$T \in \underbrace{V \otimes \cdots \otimes V}_{p \text{ copies}} \otimes \underbrace{V^* \otimes \cdots \otimes V^*}_{q \text{ copies}}$$

See e.g. [9] for details of the standard indicial formulation of tensors and results of tensor calculus.

As noted above, there is no source of objective randomness in the current formalism. Uncertainty and prediction error—and hence, VFE—is generated in the current formalism by S's in-principle ignorance of both the state $|E\rangle$ and between-observations dynamics $\mathcal{P}_E(t_U)$ of its environment. As the bits S reads from $\mathcal{B}$ are written by $\mathcal{P}_E$, S's ability to predict the future states of its informative sectors, and hence to minimize Er_X for each QRF X via Eq. (6.8), depends on its ability to predict the behavior of $\mathcal{P}_E$ locally on each informative sector. This is the task of its generative model. As the thermodynamic sector Θ is not observed, direct predictions on Θ are not possible; the local behavior of $\mathcal{P}_E$ on Θ can at best be predicted from its local behavior elsewhere. An animal, for example, must employ its available senses—hence its observable sectors—to predict the nutritional value of food.

The option space governing S's ability to locally predict $\mathcal{P}_E$ is summarized in Fig. 6.1. What is important for S is not the dynamic complexity or even the dimension of $\mathcal{P}_E$, both of which are unobservable in principle, but rather the dynamic complexity of the action of $\mathcal{P}_E$ on $\mathscr{B}$ (the dimension of this action is, clearly, just the dimension of $\mathscr{B}$). Here, the weak-interaction limit that allows separability between S and E is significant: H_{SE} (and hence $\mathscr{B}$) must have significantly lower dimension that H_E (and hence $\mathcal{P}_E$) if the weak interaction limit is to hold. The simplest case is shown in Fig. 6.1, Panel (a), in which the system E is a trivial agent deploying no QRFs other than the choice of basis for interactions with $\mathscr{B}$. The action of $\mathcal{P}_E$ is, in this case, limited to choice of basis; basis rotation by E generates quantum noise in the communication channel defined by H_{SE} that is indistinguishable by S from classical noise. Hence, the trivial agent E in Fig. 6.1, Panel (a) "looks like" a noise source to S. Emission of Hawking radiation from a BH provides perhaps the most pure example of such a noise source; while the dimension "inside" the BH can be arbitrarily large (see e.g. [11, Fig. 19]), the internal dynamics are uncoupled from the classical information encoded on the horizon and hence have no classical computational power. As will be discussed in Sect. 6.2 below, a "small" trivial agent will be driven by the FEP toward entanglement with the larger system S.

The more interesting parts of S's option space for prediction are shown in Fig. 6.1, Panel (b), (c), and (d), in which E is nontrivial. If E is nontrivial, it deploys at least one QRF X_E acting on a sector X_E. If X_E does not overlap any *informative* sector for S, however, E will appear trivial, i.e. as a noise source, to S. Hence, the interesting cases are the ones in which S's and E's informative sectors overlap; this is the case, intuitively, in which S and E can "see each other" and hence interact in the ordinary, nontechnical sense of that term.

In Fig. 6.1, Panel (b), E's sector X_E fully contains X_S. Preparation by E of the bits in X_S will, therefore, in general depend on bits outside of X_S but within X_E, i.e. on bits on the remainder $X_E \setminus X_S$. The values of these bits are "nonlocal hidden variables" [12] from S's perspective; they affect what is observed on sector X_S without being local to, i.e., contained within, X_S. Indeed, such bits may be within Θ and hence unobservable in principle by S. Changes in the values of these nonlocal hidden variables are, effectively, context changes as discussed in Chap. 5; probability distributions $P(X_S|\zeta)$ and $P(X_S|\xi)$ for distinct hidden-variable states ζ and ξ may be different. The "context-blind" distribution $P(X_S)$ can, in this case, fail to be well-defined over time. Such failures manifest as violations of Leggett-Garg inequalities [13], i.e. as "quantum hysteresis" effects due to nonlocal (i.e. outside of X_S) and possibly unobservable causes. They appear as failures to solve the Frame Problem as discussed in Box 5.2. If X_S is a reference sector for S, Frame Problem solution failures on X_A can result in failures of object re-identification [14].

A situation in which X_S fully contains X_E presents similar issues to that in Fig. 6.1, Panel (b), except here the "hidden variables" are in $X_S \setminus X_E$ and hence are accessible to S. The bits in $X_S \setminus X_E$ nonetheless contribute VFE—effectively, noise—to X_S that is unconstrained by X_E. If X_S and X_E overlap with remainders, as in Fig. 6.1, Panel (c), a similar noise contribution to X_E (or on E's side, to X_E) results. The final possibility is, clearly, that in which $X_S = X_E$ as shown in Fig. 6.1, Panel (d). Here, the source of VFE is not noise, but rather differences in the computations implemented by the QRFs X_S and X_E.

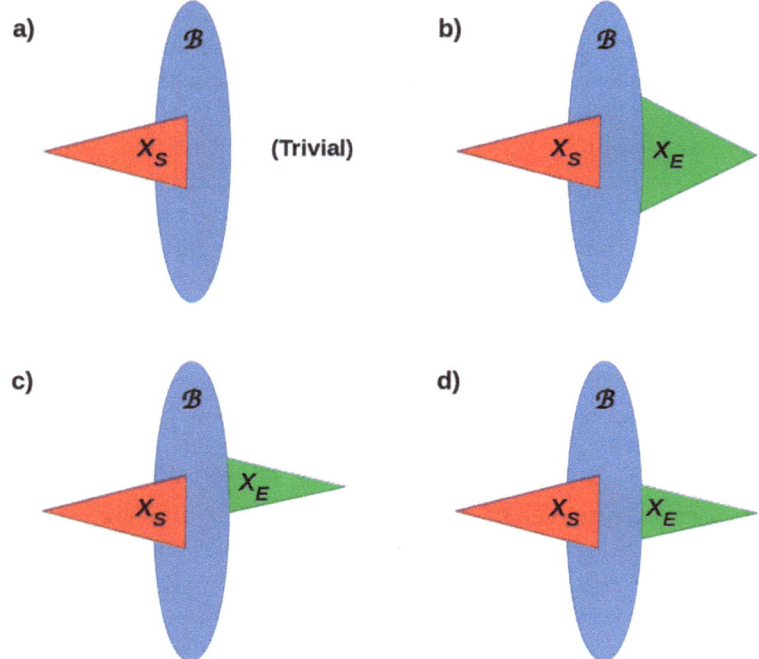

Fig. 6.1 Four options for S's ability to predict the local behavior of $\mathcal{P}_E$ on an observable sector X_S. **a** A trivial agent deploying no QRFs beyond choice of basis for interacting with $\mathcal{B}$ appears as a noise source to S. **b** E encodes a sector X_E that contains X_S; the bits on X_E but outside X_S encode "nonlocal hidden variables" for S. **c** The sectors X_S and X_E overlap; the areas of non-overlap become noise sources. **d** If $X_S = X_E$, VFE is generated by insufficient learning. Adapted from [1] Fig. 6; used with permission

Such differences correspond to differences in the structures of the CCCDs implementing the respective QRFs, e.g. differences in the "connection weights" if these are thought of as ANNs or VAEs. They correspond, in other words, to learning failures, e.g. due to insufficient training-set representativeness, as Eq. (6.6) renders obvious. We can, therefore, represent the overall situation for any observer A as:

$$\text{VFE} \; = \; \text{Noise} \; + \; \text{Insufficient Learning} \tag{6.9}$$

consistently with Eq. (4.1) above. Here "noise" includes VFE generated by unobserved context changes (Leggett-Garg violations or Frame Problem solution failures) as well as "classical" noise.

6.2   Asymptotic Behavior Under the qFEP

Classically, the FEP drives interacting systems to best-possible mutual predictability, which for spatially-separated systems manifests as generalized synchrony [2, 8]. In the current, quantum framework, optimal prediction corresponds, in the notation of Eq. (6.8), to $Er_X(k) \to 0$ as $k \to \infty$ for all observable sectors X. This clearly involves implementing a learning operator $\mathcal{L}_X$ for each sector X that is capable of asymptotically-perfect learning: let us assume this is the case. What remains given Eq. (6.9) is noise, including noise due to observed context switches. Only one mechanism for removing noise is available: that shown in Fig. 6.1, Panel (d). Hence we can conclude:

The qFEP asymptotically drives alignment of QRFs across $\mathcal{B}$.

It drives any observer S, in particular, to match any context switches by its interaction partner E in order to maintain QRF alignment. Let us now consider, therefore, a situation in which *all* QRFs deployed by S and E are aligned as in Fig. 6.1d, and in which *all* of S's QRFs have learned the local behavior of $\mathcal{P}_E$ on their sectors perfectly. The local behavior of $\mathcal{P}_E$ on some shared sector X is determined by E's QRF acting on that sector. This QRF, which we can also call X_E is, however, a quantum computation; as such, it encodes nonfungible—not finitely classically encodable—information as shown in [15]. The future behavior of X_E can, therefore, only be perfectly predicted by X_E itself, that is:

$$Er_X \to 0 \;\Rightarrow\; X_S = X_E \qquad\qquad (6.10)$$

If S and E are separable, the consequent in Eq. (6.10) violates the no-cloning theorem [16]: it demands that the internal quantum state $|E|_X(t)\rangle$, the time (external t) evolution of which implements X_E, be replicated exactly in S. Hence, if Eq. (6.10) holds, S and E cannot be separable. Therefore we have:

If $U = SE$ is isolated, the qFEP asymptotically drives the joint state $|SE\rangle$ to entanglement.

That unitary evolution drives all isolated systems to entanglement, i.e. that separability is a special case, an approximation that holds only under conditions of weak interactions and short observation times, is axiomatic in quantum theory. Hence we have:

The qFEP is, asymptotically, the Principle of Unitarity.

Any quantum system, therefore, *must* behave in accord with the qFEP; doing so is simply approaching entanglement with its environment as required by the Principle of Unitarity.

6.3   The qFEP as a Scale-Free Theory

Quantum theory originated as a theory of atomic-scale phenomena, and is still often presented as relevant only in domains in which the magnitudes of typical actions are on the order of $\hbar$. As mentioned earlier, the classical FEP treats information flow as continuous; from this perspective, quantum theory becomes relevant only when it is no longer possible to treat fluctuations from averages as classical noise [8].

Information, however, is quantized at all scales. Yes/no questions exist at all scales, and the appropriateness of actions depends on their discrete answers at all scales. This is most obvious, at our human scale, in the case of questions about individual identity, i.e. about the persistence through time of particular systems. Is this *my* laptop? If so, particular actions are available. If not, while some of those actions remain possible, they may have very different consequences. This basic logic applies to "classical" objects across the board. What systems can be said to "know" in some cases applies to whole classes of things, but in many of the most important cases applies only to specific, sometimes difficult-to-identify individuals.

The fundamental discreteness of information renders quantum theory scale free, and hence renders the qFEP scale free. The qFEP describes all systems, regardless of scale, as engaging in bidirectional communication with their environments at all times. The goal of this communication, in every case, is the preservation of the system-environment distinction, i.e. the preservation of a separable joint state $|SE\rangle$. Asymptotically, however, the process of error minimization achieves exactly the opposite goal: system-environment entanglement. This asymptotic limit, like quantum theory itself, is fully scale free.

References

1. Fields, C.; Friston, K.J.; Glazebrook J.F.; Levin M. A free energy principle for generic quantum systems. Prog. Biophys. Mol. Biol. 2022, 173, 36–59.
2. Fields, C.; Fabrocini, Friston, K.; Glazebrook, J.F.; Hazan, H.; Levin, L.; Marcianò, A. Control flow in active inference systems, Part I: Formulations of classical and quantum active inference. IEEE Trans. Mol. Biol. Multi-Scale Commun. 2023, 9, 235–245.
3. Fields, C., Fabrocini, F., Friston, K., Glazebrook, J. F., Hazan, H., Levin, M. and Marcianò, A. (2023) Control flow in active inference systems, Part II: Tensor networks as general models of control flow. IEEE Transactions on Molecular, Biological, and Multi-Scale Communications 9: 246-256
4. Fields, C., Friston, K., Glazebrook, J. F., Levin, M. and Marcianò, A. (2022) The free energy principle induces neuromorphic development. Neuromorphic Computing and Engineering 2: 042002
5. Fields, C. and Levin, M. (2023) Regulative development as a model for origin of life and artificial life studies. BioSystems 229: 104927
6. Fields, C. (2024) The free energy principle induces compartmentalization. Biochemical and Biophysical Research Communications 723: 150070.
7. Fields, C.; Goldstein, A.; Sandved-Smith, L. Making the thermodynamic cost of active inference explicit. Entropy 2024; 26, 622.

8. Friston, K.J. A free energy principle for a particular physics. arXiv 2019, arXiv:1906.10184.

9. Abraham, R., Marsden, J. and Ratiu, T. *Manifolds, Tensor Analysiis and Applications*. Appl. Math. Sciences 75 (1988), Springer, New York.

10. Orús, R. (2019). Tensor networks for complex quantum systems. *Nat. Rev. Phys.* 1, 538–550.

11. Almheiri A, Hartman T, Maldacena J, Shaghoulian E, Tajdini A. The entropy of Hawking radiation. Rev Mod Phys 2021;93:035002.

12. Mermin, N. D. (1993). Hidden variables and the two theorems of John Bell. Reviews of Modern Physics 65, 803–815.

13. Emary C, Lambert N, Nori F. Leggett-Garg inequalities. Rep Prog Phys 2013;77:016001.

14. Fields, C. (2013) How humans solve the frame problem. Journal of Experimental and Theoretical Artificial Intelligence 25: 441–456.

15. Bartlett, S.D.; Rudolph, T.; Spekkens, R.W. Reference frames, superselection rules, and quantum information. Rev. Mod. Phys. 2007, 79: 555–609.

16. Wootters, W. T.; Zurek, W. H. A single quantum cannot be cloned, Nature 1982; 299: 802–803.

Multi-agent Communication, Games, and Computational Complexity

7

Recall from Chap. 2, Sect. 2.6 that two or more agents can communicate using LOCC protocols [1], and that any quantum system accessed by multiple observers—an apparatus, or even a shared environment—can serve as a quantum channel. Here we develop these ideas with greater rigor, employing the formalism of *topological quantum field theories* (TQFTs) [2, 3], a generic, minimal assumption formalism for describing the unitary evolution of isolated systems through an assumed background time, i.e. the t_U introduced in Chap. 2. Formally, a TQFT is a functor (see Appendix A for definition) from the category **Cob** of cobordisms to the category **Vect** of vector spaces. A cobordism is a manifold of dimension $n + 1$ whose boundary is a union of manifolds of dimension n; for example, a cylinder with two disks as its boundary. A TQFT is an assignment of vector spaces to the manifolds composing the boundary, and linear maps, i.e. vector-space homomorphisms, to the manifold enclosed by the boundary. In the TQFTs of interest here, the boundary vector spaces are Hilbert spaces of some quantum system S, and the manifold of maps comprises all components—all possible paths—by which S can evolve through time. The manifold of maps is often called the "bulk"; we will use this terminology here. A great advantage of the TQFT formalism is its lack of any "spatial" assumptions beyond that of native metric-space structure inherited from **Vect**; hence it is a natural formalism for describing the emergence of spacetimes from quantum dynamics in some background time t_U.

© The Author(s), under exclusive license to Springer Nature Switzerland AG 2026

C. Fields and J. Glazebrook, *Distributed Information and Computation in Generic Quantum Systems*, Synthesis Lectures on Engineering, Science, and Technology, https://doi.org/10.1007/978-3-031-97263-8_7

7.1   Multiple Components as Multiple Agents

7.1.1   Sequential Measurements

We first show that sequential observations of any finite physical system S that employ either one or some sequence of QRFs induce a TQFT on S; we follow the development and use the notation of [4]. We start with a functor between categories, $\mathfrak{F} : \mathbf{CCCD} \longrightarrow \mathbf{Cob}$, the former the category of CCCDs representing QRFs, and the latter the category of finite cobordisms. The morphisms in $\mathbf{CCCD}$ produce arbitrarily large structures computing arbitrarily complex Boolean functions that represent arbitrarily large, multi-variate QRFs. This principle is applied to sequential measurements of S, which we can also view as a sector S of some boundary $\mathscr{B}$. We assume that the measurements employ a single QRF, which for convenience we also denote S. We can then consider factoring the system S by factoring the QRF S; we use triangles to represent the relevant CCCDs:

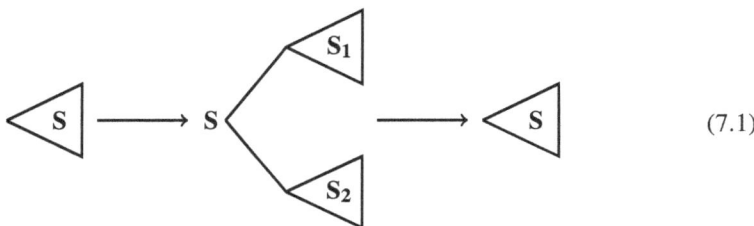

$$(7.1)$$

Diagram (7.1) is a straightforward relabelling of subsets of m base-level classifiers within a CCCD as in [4, Sect. 3]:

$$\underbrace{\mathcal{A}_1, \mathcal{A}_2, \ldots \mathcal{A}_m}_{S} \rightarrow \underbrace{\mathcal{A}_1, \ldots \mathcal{A}_i,}_{S_1} \underbrace{\mathcal{A}_{i+1}, \ldots \mathcal{A}_m}_{S_2} \rightarrow \underbrace{\mathcal{A}_1, \mathcal{A}_2, \ldots \mathcal{A}_m}_{S} \qquad (7.2)$$

that permit defining limits/colimits $\mathbf{S_1}$, and $\mathbf{S_2}$ over the $\mathcal{A}_1, \ldots \mathcal{A}_i$ and the $\mathcal{A}_{i+1}, \ldots \mathcal{A}_m$, respectively. Because the limit/colimit $\mathbf{S}$ exists over the $\mathcal{A}_1, \mathcal{A}_2, \ldots \mathcal{A}_m$, it clearly exists over $\mathbf{S_1}$ and $\mathbf{S_2}$.

Next, we take the initial and final measurements to involve only the reference component R of S. The pointer component P is traced over in these measurements. These reference measurements serve to confirm the identity of S and hence assure that the subsequent measurements are indeed sequential measurements of one and the same system. This second sequence is represented as:

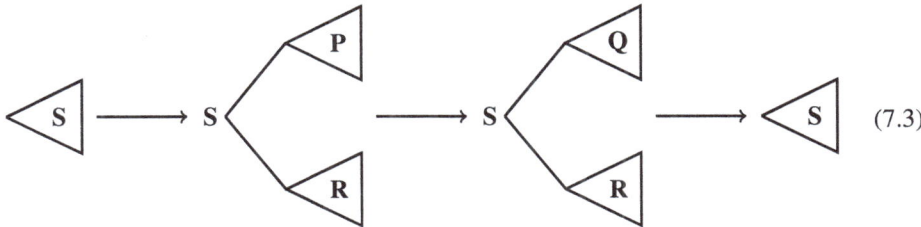

Letting $k < m$, we can represent this in terms of classifier labels, leaving traced-over classifiers implicit, as:

$$\underbrace{\mathcal{A}_1, \mathcal{A}_2, \ldots \mathcal{A}_k}_{R} \rightarrow \underbrace{\mathcal{A}_1, \ldots \mathcal{A}_k,}_{R} \underbrace{\mathcal{A}_{k+1}, \ldots \mathcal{A}_m}_{P} \rightarrow \underbrace{\mathcal{A}_1, \ldots \mathcal{A}_k,}_{R} \underbrace{\tilde{\mathcal{A}}_{k+1}, \ldots \tilde{\mathcal{A}}_m}_{Q} \rightarrow \underbrace{\mathcal{A}_1, \mathcal{A}_2, \ldots \mathcal{A}_k}_{R} \quad (7.4)$$

where the notation $\tilde{\mathcal{A}}_l$ indicates that $\mathcal{A}_l$ has been rewritten in a rotated measurement basis, e.g. $s_z \rightarrow s_x$ or $x \rightarrow p = m \, (\partial x / \partial t)$. As with Eqs. (7.2), (7.4) is a relabeling, but here the local basis rotation mapping $\mathbf{P} \rightarrow \mathbf{Q}$ has been added. As both $\mathbf{P}$ and $\mathbf{Q}$ must commute with $\mathbf{R}$, the commutativity requirements for $\mathbf{S}$ are satisfied.

As shown in [4, Sect. 3], these can be mapped to TQFTs. For the first case:

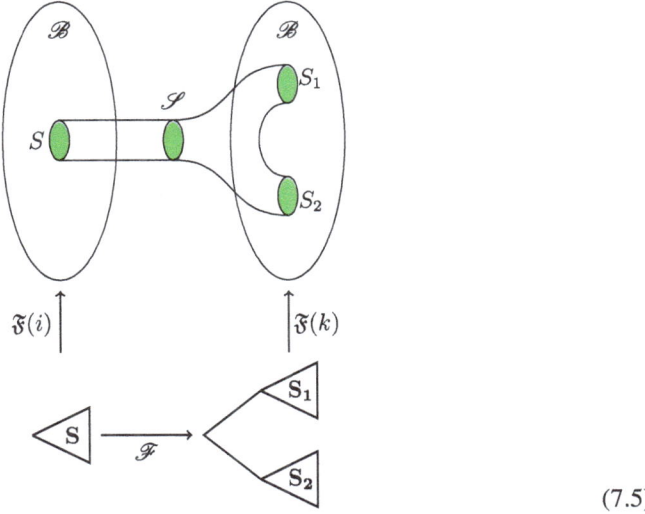

$$(7.5)$$

In the second case, the initial and final measurements are of the reference component R only; the pointer component is traced over. Hence we have:

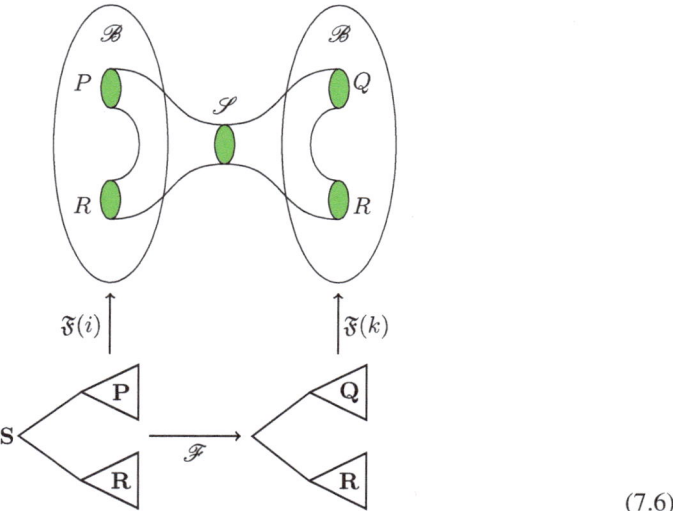

$$(7.6)$$

Diagrams (7.5) and (7.6) illustrate the compatibility between objects (representing QRFs) and arrows (representing sequential measurements) in **CCCD**, and the resulting cobordisms on sectors, both respecting composition and identities, and in particular, an explicit representation of the functor $\mathfrak{F}$. In fact, for any morphism $\mathscr{F}$ of CCCDs in **CCCD**, there is a cobordism $\mathscr{S}$ such that a diagram of the form of Diagram (7.5) or (7.6) commutes [4, Theorem 2].

We remark that a property of $\mathfrak{F}$ is that any sequence of actions of, or equivalently, operations with one or more QRFs, can simply be identified with the TQFT that it induces. This is quite natural since a QRF is a physical system Q with some internal dynamics representable by a Hamiltonian H_Q. Hence treating it as isolated, it evolves in time t via a unitary operator $\mathcal{P}_Q = \exp[-(\iota/\hbar)H_Q t]$. The "boundaries" between which Q can be considered isolated are precisely its actions on S at some sequential times $t_i, t_j, \ldots t_k$. Evolving S through time and evolving Q through time are, therefore, operationally the same process: they yield precisely the same data, the data obtained by acting with Q on S at $t_i, t_j, \ldots t_k$. Significantly, this amounts to a *bulk-boundary duality*, with Q as the bulk, and $\mathcal{H}_S$ as the boundary [4, 5].

7.1.2   QRF Sharing Induces Entanglement

The identification of QRFs with the TQFTs can be used to construct a novel, and purely topological proof that QRF sharing induces entanglement, a result previously demonstrated via the no-cloning theorem [6, 7]. As QRFs are by definition physical systems, their quantum states encode unmeasurable quantum phase information (in [8] this is called "nonfungible information"). The state—in particular, the initial or "ready" state—of a QRF cannot, therefore, be determined by any finite number of finite-resolution observations. Whether, in particular, the prepared state of a QRF is identical to a theoretically-specified state cannot be determined by finite observations. Unknown quantum states cannot, in general, be cloned by any unitary process [9]. A practical consequence is that QRF sharing required by quantum communication protocols (for instance, sharing an independently-measurable z axis) can only be approximate. Proving this [10] follows some basic groundwork as below.

Recall the consequences of imposing separability from Chap. 2. Let $U = AB$ be a bipartite decomposition of a closed system U that satisfies the separability constraint $|AB\rangle = |A\rangle|B\rangle$ over any time period of interest; this can be achieved provided the interaction H_{AB} is sufficiently weak. Under these conditions, the decompositional boundary functions as a holographic screen $\mathscr{B}$. This $\mathscr{B}$ is a discrete topological space comprising N mutually-disjoint sites, where $2^N = \dim(H_{AB})$. As illustrated in Fig. 2.1, these N sites can be considered to each house a single qubit, with the N qubits collectively encoding the 2^N eigenvalues of H_{AB} [4, 6, 11–13]. The two systems A and B can be considered as "agents"—indeed FEP-compliant agents—Alice and Bob that communicate by exchanging messages across $\mathscr{B}$; formally (and often in practice), we can consider them to alternately *prepare*, and then *measure* the qubits on $\mathscr{B}$. This situation is illustrated in Diagram (7.7). Note that as the Hilbert space $\mathcal{H}_U = \mathcal{H}_A \otimes \mathcal{H}_B$, these N qubits are ancillary to $\mathcal{H}_U$. and to the self-interaction H_U.

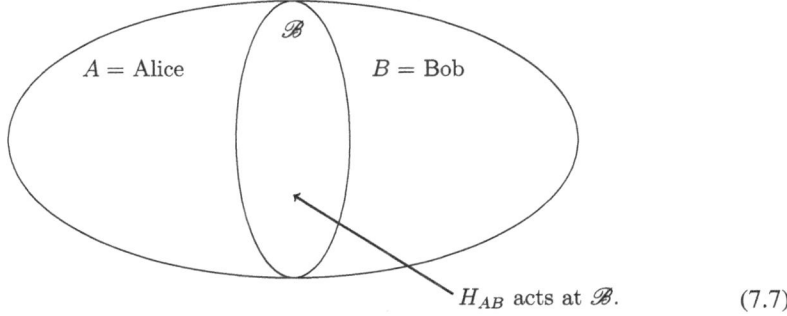

$$H_{AB} \text{ acts at } \mathscr{B}. \tag{7.7}$$

Suppose now that A and B operate on $\mathscr{B}$ with QRFs Q_m and Q'_m, respectively; here we follow [4] in using the notation 'Q' to denote both a QRF, as well as the dual preparation and measurement operations performed using that QRF in question. The domains $\dom(Q_m)$ and $\dom(Q'_m)$ are subsets of qubits that encode the "states" of measured (dually, prepared) "systems" (or "messages") S and S', respectively. Suppose further that A and B record their observational outcomes on $\mathscr{B}$ with QRFs Q_r and Q'_r, respectively; the domains $\dom(Q_r)$ and $\dom(Q'_r)$ are subsets of qubits that function as "memories" Y and Y', respectively. For convenience, we will assume $\dim(S) = \dim(Y)$, and $\dim(S') = \dim(Y')$.

The composite operation $Q = (\overrightarrow{Q}, \overleftarrow{Q})$, where $\overrightarrow{Q} = Q_r Q_m$ and $\overleftarrow{Q} = Q_m Q_r$, is then a pair of QRF sequences that can be identified with TQFTs that measure and record an outcome, mapping $\mathcal{H}_S \to \mathcal{H}_Y$, and dually use an outcome read from memory to prepare a state, mapping $\mathcal{H}_Y \to \mathcal{H}_S$, respectively; the composite operation Q' can be defined similarly. These composite operations strongly demonstrate Wheeler's contention (and indeed that of Niels Bohr), that a quantum measurement cannot be considered to have occurred until it is irreversibly recorded [14], in this case, written to the memory Y. This process can be represented as:

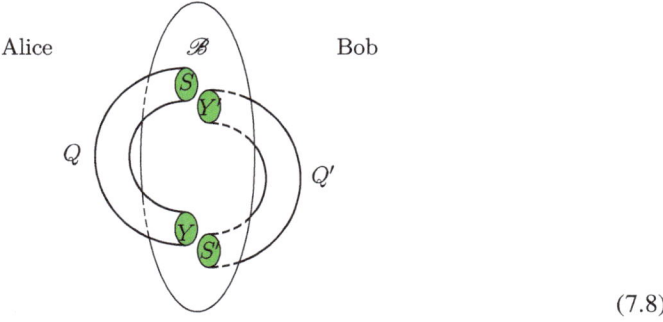

$$(7.8)$$

where Alice and Bob are represented as the components of the bulk to the left and right of $\mathscr{B}$, respectively. In this Diagram (7.8), Alice's and Bob's QRFs access different systems S and S', as well as different memories Y and Y', and can be considered completely independent. They are, in particular, conditionally independent when $\mathscr{B}$ is replaced with a classical Markov blanket (MB), [15]—see [7] for details.[1]

Now consider the case in which $S = Y'$ and $Y = S'$:

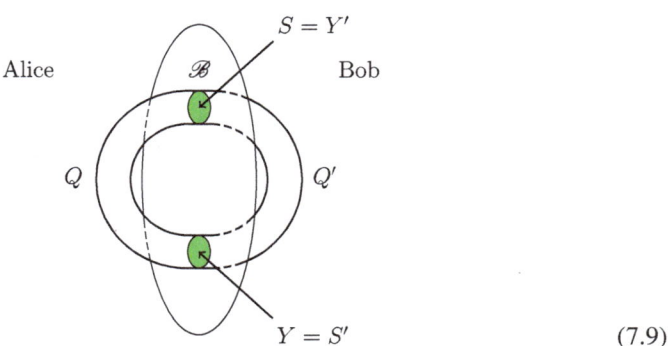

$$(7.9)$$

---

[1] Viewing $\mathscr{B}$ as a holographic screen, to see how it functions as an MB separating A from B, we regard $\mathscr{B}$ as having an 2^N-dimensional, N-qubit Hilbert space $\mathcal{H}_{q_i} = \prod_i q_i$. While $\mathcal{H}_{q_i}$ is strictly ancillary to $\mathcal{H}_U = \mathcal{H}_A \otimes \mathcal{H}_B$, the classical situation can be recovered in the limit in which the entanglement entropies $\mathcal{S}(|A\rangle), \mathcal{S}(|B\rangle) \to 0$ by considering the products $\mathcal{H}_A \otimes \mathcal{H}_{q_i}$ and $\mathcal{H}_B \otimes \mathcal{H}_{q_i}$ to be "particle" state spaces for A and B, respectively, for which, in this classical limit, the states of $\mathcal{H}_{q_i}$ then become the blanket states of an MB, which is trained inductively by conditional independence [4, 12].

There are two consistent temporal flows in Diagram (7.9): $\overrightarrow{Q}\,\overrightarrow{Q'}$ and $\overleftarrow{Q'}\,\overleftarrow{Q}$, corresponding to each observer reading the other's memory and each observer measuring the other's system, respectively. Both $\overrightarrow{Q}\,\overrightarrow{Q'}$ and $\overleftarrow{Q'}\,\overleftarrow{Q}$ are, however, TQFTs with copies of S as boundaries. By sharing QRF's, therefore, Alice and Bob jointly implement a single unitary process. As a consequence, they are entangled.

Diagram (7.9) provides a new perspective on the no-cloning theorem, as it states that a unitary process cannot be "cloned" by separable observers. When a process is viewed as an operator applied to a state, it becomes clear why this must be the case. As noted above, neither an unknown quantum state, nor an unknown unitary operator, can be fully specified by a finite set of finite-resolution observational outcomes. While Alice and Bob may share *a priori* theoretical specifications of the states to be prepared, the fact is, whether the prepared states meet these specifications exactly cannot be determined by finite observation. All that Alice and Bob can learn about each other's state preparations is contained in the finite sets of observational outcomes that each can obtain by interacting with $\mathscr{B}$. Specifically, they can obtain only the set of eigenvalues of H_{AB} encoded at finite resolution on $\mathscr{B}$. These observed eigenvalues are insufficient, in particular, to determine the entanglement entropy $\mathcal{S}(|AB\rangle)$ of the observers' joint state; see [16] Corollary 3.1 for proof. Alice and Bob cannot, therefore, determine by observation whether they are entangled or mutually separable.

7.1.3   Multi-party Communication Protocols

Let us commence by remarking that separated observers cannot share QRFs; observational outcomes are observer-relative not as a matter of interpretation, but rather in principle [5]. Hence quantum theory is naturally formulated as a single-observer theory [17–23]. In practice, however, we are mainly interested in situations in which two or more observers share an environment, including its information-processing resources, whether quantum or classical. In such practical settings we require, moreover, the observers in question to be separable, i.e. that their joint state is separable at all times.

Let us suppose, then, that Alice is a bipartite system comprising two observers, A_1 and A_2, who interact with a shared environment, Bob. We further suppose that A_1 and A_2 deploy composite QRFs Q_1 and Q_2. Extending this decomposition to multiple observers is a simple, iterative process, but is hard on notation and graphically cumbersome; the important results can be obtained considering just two observers.

As discussed in Chap. 5, the QRFs Q_1 and Q_2 commute if and only if their representations as CCCDs commute, i.e. if and only if their representations as CCCDs have a mutual limit and a mutual colimit [4]. Non-commutativity of QRFs induces quantum contextuality [6]—see Chap. 5, [16, Theorem 3.4] and [24, Sect. 7] for details.

Here we show that if Q_1 and Q_2 commute, A_1 and A_2 are entangled, and so cannot be considered distinct observers. This immediately implies that *all* observations recorded by distinct, i.e. mutually-separable observers are context-dependent. We first demonstrate this result and show how it applies in a canonical case of joint manipulation of entanglement as a resource, then show how decoherence explains the observational elusiveness of the implied contextuality.

We first consider the case in which Q_1 and Q_2 have distinct pairs of domains S_1 and Y_1 and S_2 and Y_2, respectively. In this case, nothing is changed by considering the disjoint unions $S_1 \sqcup Y_1$ and $S_2 \sqcup Y_2$ as the domains of Q_1 and Q_2, respectively. Hence we have:

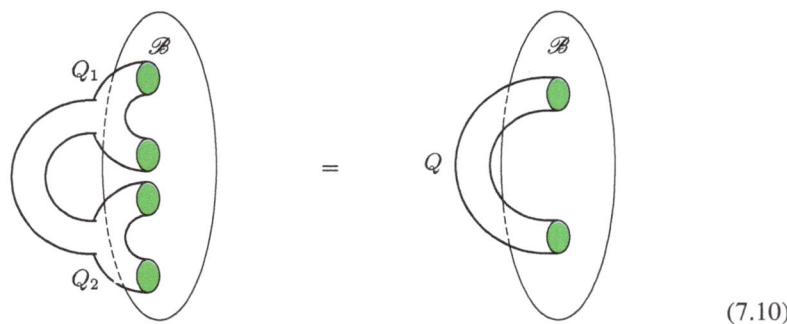

$$(7.10)$$

Here we make no assumptions about QRFs deployed by Bob, but treat H_B as implementing a generic quantum process. The right-hand side of Diagram (7.10) depicts a single TQFT, and hence a unitary process, implemented by Alice. A_1 and A_2 are, therefore, entangled. An explicit construction of this result employing the CCCD representation of the relevant QRFs and the algebraic operation of "concurrence" on CCCDs is provided in [4, Appendix A.5] and the concurrence operation provides a general model of computational concurrency with a defined semantics, as summarized in [4, Appendix A.2].

The second case of interest is that in which the domains of Q_1 and Q_2 are not independent, but rather overlap. Suppose $S_1 = S_2 = S$; all other domain-overlap cases are similar. Commutativity of Q_1 and Q_2 implies that A_1 can prepare S simultaneously with A_2's measurement of S. Here again, A_1 and A_2 share a single unitary process and are, therefore, entangled. As in the case of Diagram (7.9), A_1 and A_2 cannot measure their joint-state entanglement entropy; hence they cannot determine by observation that they are entangled, and hence cannot determine by observations that Q_1 and Q_2 commute.

Maintaining the distinction, i.e. separability, between A_1 and A_2 requires, therefore, that Q_1 and Q_2 do not commute. In this case, A_2's outcome probabilities for measurements of S_2 depend on A_1's preparations of S_1 and vice-versa. This is contextuality by default, not as an operational assumption [25], but as a first-principles requirement—cf. Chap. 5, [26] and [24, Sect. 7] [4, Sect. 7.2]. Contextuality is, however, difficult to observe in practice, as reviewed in [27, 28]. Diagram (7.7) makes it clear why: over the extended cycles of measurement required to accumulate statistics, Bob's internal dynamics H_B mixes pure states $|S_1\rangle$ and $|S_2\rangle$ to decoherent densities ρ_{S_1} and ρ_{S_2}. A_1 and A_2 do not, in general, experience contextuality because they cannot, in general, measure a sufficiently large fraction of $\mathcal{B}$.

7.2 LOCC Protocols

7.2.1 Defining LOCCs

All communication protocols assume distinct, i.e. separable agents/observers capable of exchanging classical information via a shared language. As a language ability can be taken to be implemented by a QRF [6, 7], the results of Sect. 7.1.2 require that separable agents can at most approximately share a language. Classical communication is enacted by encoding messages into and decoding messages from some physical medium, e.g. the ambient photon field. Such media are quantum systems comprising degrees of freedom of the environment shared by the communicating agents (cf. the discussion in [29]); in the present notation, they are components of Bob. A medium implementing an optimal, noise-free classical channel can be represented as a QRF implemented by Bob. Such a channel introduces, at most, a basis rotation of the encoded data; such a basis rotation is sufficient to assure non-commutativity of the sending and receiving QRFs. Treating the message space as a shared classical memory, we can represent this channel QRF as a TQFT mapping $Y_1 \leftrightarrows Y_2$. Hence we can represent classical communication by:

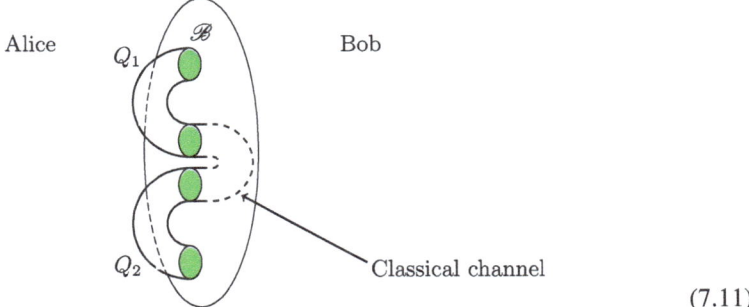

$$(7.11)$$

where A_1 and A_2 are assumed to be distinct agents and hence Q_1 and Q_2 are assumed not to commute. Note that from an operational perspective, this assumption of non-commutativity corresponds to the non-observability of the channel QRF implemented by Bob, which is separated from A_1 and A_2 by $\mathscr{B}$. A_1 and A_2 cannot, therefore, deduce the preparation conditions that this channel imposes on Y_2, or in the reverse direction, on Y_1. This is in fact the case in ordinary classical communication between humans, and is why classical eavesdroppers can avoid detection. From a classical perspective, the boundary $\mathscr{B}$ implements an MB separating the observers from the channel and restricting their access to it [4, 7].

The concept of a LOCC protocol, introduced in Chap. 2, Sect. 2.6, follows from how quantum instruments and quantum channels are broadly specified in terms of one-to-one correspondences between these and *trace-preserving, completely positive* (TCP) maps of bounded operators on Hilbert spaces; we refer to Chitambar et al. [1] for the technical details. An instrument $\mathfrak{I}'$ is said to be *LOCC-linked* to $\mathfrak{I}$, if $\mathfrak{I}'$ is a coarse-graining of $\mathfrak{I}$. By "coarse-graining", we mean a procedure of grouping instruments in $\mathfrak{I}$ in a compatible way

[1]. Then, LOCCs are defined recursively in the following way: we say that $\mathfrak{I}$ is LOCC$_1$ if it is local with respect to some party. Then we say that $\mathfrak{I}$ is LOCC$_n$ if it is LOCC linked to some $\mathfrak{J}$ which is LOCC$_{n-1}$. LOCC is then defined as the direct limit of a system of LOCC$_n$ instruments. For much of our discussion, only two-observer, i.e. LOCC$_2$ protocols need be considered.

In a LOCC protocol, A_1 and A_2 are assumed to know the preparation conditions for their shared quantum channel; indeed these preparation conditions are what they classically communicate. Hence we can represent a LOCC protocol by adding a quantum channel—a TQFT—to Diagram (7.11) to produce Diagram (2.4), copied below for convenience:

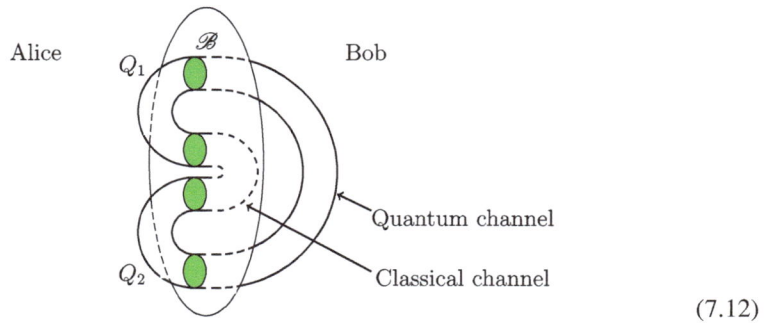

$$(7.12)$$

Note that the two channel QRFs implemented by Bob in Diagram (7.12) are functionally equivalent: the labels "classical" and "quantum" can be exchanged without altering the diagram. This symmetry reveals the dependence of LOCC protocols on the assumption that communication between observers in language can be regarded as classical. This assumption of classicality operationalizes the assumption of separability between observers that is required for the notions of "joint observation" and "joint manipulation" to be physically meaningful. Note also that the intersections of the classical information channels with $\mathscr{B}$ in (7.11) and (7.12), comprise the (classical) MBs of Alice and Bob, respectively.

7.2.2   Examples

Example 7.2.1   Entanglement of formation as a measure of non-commutativity Protocols that involve the manipulation of multiple entangled states provide a more elaborate example of LOCC that single-qubit exchange. Wootters [30] considers the case in which two observers jointly prepare a shared mixed state ρ using shared singlet states as a resource. Classical communication is clearly required; it is assumed that no additional quantum communication channel is available. Adopting the notation of [30] in which some set $\{\Psi_i\}$ of singlet states are shared by parties A and B, the number of shared singlets that must be consumed to create n copies of some given pure state $|\Phi\rangle$ is $nE(\Phi)$, where $E(\Phi)$ is the von Neumann entanglement entropy $\mathcal{S}$ across the A–B partition [30, Eq. 3]:

$$E(\Phi) = \mathcal{S}(\operatorname{tr}_B |\Phi\rangle\langle\Phi|) = \mathcal{S}(\operatorname{tr}_A |\Phi\rangle\langle\Phi|) = -\sum_i^n c_i^2 \log_2 c_i^2 \qquad (7.13)$$

where the c_i are Schmidt coefficients. Suppose now that the goal is to create n copies of a mixed state ρ that can be decomposed into pure states as [30, Eq. 4]:

$$\rho = \sum_{j=1}^N p_j |\Phi_j\rangle\langle\Phi_j| \qquad (7.14)$$

Here the $|\Phi_j\rangle$ are distinct, but not necessarily orthogonal, normalized pure states of the bipartite (A, B) system, and $p_j \geq 0$ (with $\sum_j p_j = 1$). The number of singlets expended is [30, Eq. 5]:

$$n \sum_{j=1}^N p_j E(\Phi_j) \qquad (7.15)$$

with $E(\Phi_j)$ given by Eq. (7.13).

The *entanglement of formation* of the mixed state ρ is then defined by [30, Eq. 6]:

$$E_f(\rho) = \inf \sum_j p_j E(\Phi_j) \qquad (7.16)$$

with the infimum taken over all pure state decompositions of ρ. Comparing Eq. (7.16) with (7.15), $E_f(\rho)$ is just the minimum number of singlets needed to make each copy of ρ.

As pointed out in [30], the entanglement of formation is not recoverable by distillation; the conversion of $\{\Psi_i\}$ to ρ is an irreversible process. The term $E_f(\rho)$ is, therefore, effectively a measure of the quantum information that this conversion renders inaccessible. To see how this information is "lost" during the conversion, it is useful to consider the process as a LOCC protocol on the model of Diagram (7.12), with A_1 and A_2 the agents performing the protocol and the state transition $\{\Psi_i\} \to \rho$ being implemented by Bob in response to their manipulations. In this setting, irreversibility corresponds to non-commutativity between Q_1 and Q_2 when interpreted as operators as discussed above. We can, therefore, write:

$$E_f(\rho) = (2/h) \|[Q_1, Q_2]\rho\| \qquad (7.17)$$

with h Planck's constant. The upper limit $E_f(\Psi) = E(\Psi) = 1$ for a singlet state corresponds to the action of fixing one unknown bit, the one not measured by the first-acting operator. The lower limit $E_f(\rho) = 0$ is achieved if and only if ρ is separable [30], i.e. only if A_1 and A_2 do not share a quantum channel traversing Bob. This is consistent with A_1 and A_2 remaining distinct agents only if the global decomposition $U = AB$ is replaced by

$U = ZW$, where $Z = \text{Alice}_1 \rho_1$, $W = \text{Alice}_2 \rho_2$, and $\rho = \rho_1 \rho_2$, exemplifying the general decomposition dependence of entanglement and separability [31–35].

Example 7.2.2 Undecidability of QRF sharing A LOCC protocol can be viewed as a method that allows A_1 and A_2 to determine, via observation and classical communication, whether they share a quantum channel; it is this perspective on LOCC that is implemented by Bell/EPR experiments [36, 37]. It can also be viewed, via the symmetry of Diagram (7.12), as allowing A_1 and A_2 to determine, via manipulations of a shared quantum resource, whether they share a classical communication channel, and hence a classical language. A "language" is a shared mapping from input data to output data, which must be physically implemented by a QRF and tracked for semantic consistency in the accompanying information flow by a CCCD. Hence we can view a LOCC protocol as allowing A_1 and A_2 to determine, via manipulations of a shared quantum resource, whether they share a QRF. As we have seen, QRF sharing induces entanglement. It is for this reason that, as noted above, A_1 and A_2 can assume that they share a common language for practical purposes, but they cannot demonstrate it by making local measurements. A_1 cannot, in particular, deduce from local measurements the QRF employed by A_2. Nor can Bob deduce from measurements that A_1 and A_2 share a QRF.

Restating this result in the language of formal undecidability reveals the deep connection between LOCC protocols and computability. Besides the interpretation in terms of measurement/preparation operations given in Sect. 7.1.2, a QRF can also be thought as an operator $Q : \psi \mapsto P_\psi(x_i)$ that acts on a quantum state ψ to yield a probability distribution $P_\psi(x_i)$ of QRF-relative outcome values x_i with well-defined units of measurement. If Q is a QRF and ψ and ϕ are quantum states, then for any unitary $U : \psi \mapsto \phi$, there must be a map $f_U : P_\psi(x_i) \mapsto P_\phi(x_i)$ such that the diagram:

$$
\begin{array}{ccc}
\psi & \overset{U}{\dashrightarrow} & \phi \\
{\scriptstyle Q}\downarrow & & \downarrow{\scriptstyle Q} \\
P_\psi(x_i) & \overset{f_U}{\longrightarrow} & P_\phi(x_i)
\end{array}
\tag{7.18}
$$

commutes, i.e. $f_U Q = QU$. Any such Q clearly respects identities and composition, i.e. if $U = VW$ there must be maps $f_U = f_V f_W$, so Q can be considered functorial. In practice, we are only interested in finite apparatus interacting with finite systems, hence we can write:

$$Q : \textbf{QuantState}_N \to \textbf{ProbDist}_N$$

with **QuantState**$_N$ the category of quantum states with some finite dimension N (i.e. rays in the Hilbert space of dimension N) and **ProbDist**$_N$ the category of probability distributions over a set of N elements. In this functorial representation, the nonfungibility of the physical system Q corresponds to the forgetfulness of the functor Q: the phase information that determines, via Q, the function f_U cannot be reconstructed from f_U itself. Indeed we can always find quantum states ζ and ξ, a unitary $V : \zeta \mapsto \xi$, and an alternative QRF $R \neq Q$ such that the diagram:

$$
\begin{array}{ccc}
\psi & \overset{U}{-\;-\;-\;-\;\rightarrow} & \phi \\[2pt]
\Big\downarrow{\scriptstyle Q} & & \Big\downarrow{\scriptstyle Q} \\[4pt]
P_{\psi,\zeta}(x_i) & \overset{f_U}{\underset{f_V}{\longrightarrow}} & P_{\phi,\xi}(x_i) \\[4pt]
\Big\uparrow{\scriptstyle R} & & \Big\uparrow{\scriptstyle R} \\[2pt]
\zeta & \underset{V}{-\;-\;-\;-\;\rightarrow} & \xi
\end{array}
\tag{7.19}
$$

commutes; here the notation $P_{a,b}(x_i)$ abbreviates $P_a(x_i) = P_b(x_i)$. From a quantum computing perspective, this is trivial: it simply says that any function f_U can be implemented to arbitrary accuracy by multiple quantum algorithms.

Let us now assume that Bob has the computational capability of a Turing machine [38, 39] with an arbitrarily large memory, and ask whether Bob can decide, given finite measurements of the states ψ, ϕ, ζ, and ξ and finite descriptions of Q and R—in the notation of Diagram (7.12), Q_1 and Q_2—whether Diagram (7.19) commutes. To make the question precise, we can assume that Q and R are described by finite programs in some Turing-complete programming language; the question is then whether Bob can decide whether $f_U = f_V$ given these two programs and a finite set of inputs. Rice's theorem [40] answers this question in the negative: no Turing machine can decide what function is computed by an arbitrarily-given program. By the same reasoning, neither A_1 nor A_2 can decide whether their respective QRFs compute the same function.

The separability between A_1 and A_2 on which LOCC protocols depend rests, therefore, on the undecidability of QRF sharing, and hence of entanglement. One of the simplest LOCC protocols is replicate measurement; Rice's theorem [40] implies that whether A_2 has replicated A_1's measurement is undecidable. In practice, confirming replication becomes increasingly difficult as the measurement resolution $N \rightarrow \infty$, i.e. as the functions f_U and f_V are specified with greater and greater precision. Replication depends, in practice, on the coarse-graining imposed by macroscopic calibration procedures, effectively-classical initial states, and finite measurement resolution. It depends, in other words, on replication being "up to" some classical noise; a point already made by Bohr in 1928 [41].

7.2.3   How Quantum Darwinism Describes a LOCC Protocol

We mentioned in Chap. 2, Sect. 2.6 that quantum Darwinism [42, 43], see [44] for a recent review, is a LOCC protocol; here we show this in detail. Classical communication is an instance of environmental redundancy: distinct observers A_1 and A_2 can share a classical message only if the message is redundantly encoded on $\mathscr{B}$ by the action of the "environmental" dynamics H_B. Redundant encoding of the quantum states of "systems" embedded in a "witnessing" environment [45, 46] is the basis of quantum Darwinism. For any system X embedded in the environment, the pointer states selected for encoding are the eigenstates of H_{XE_X}, where E_X is the environment of X, i.e. $XE_X = B$. When multiple observers A_i are present, each observer exclusively accesses a component E_{X_i} of E_X. The state $|X\rangle$ is encoded *redundantly* if each of the A_i, interacting exclusively with the single environmental component E_{X_i}, obtains the same measurement outcome, and hence reports an observation of the same state $|X\rangle$. Note that while the encoding of $|X\rangle$ in E_X is a quantum process, the observational reports that reveal the redundant encoding are classical. Redundant encoding of $|X\rangle$ in multiple components E_{X_i} of E_X clearly requires that $H_{XE_{X_i}} = H_{XE_{X_j}}$ for all components i, j. As the observers are assumed to act completely independently, they must be mutually separable; separability is preserved—hence contextuality is prevented—by requiring that the environmental sectors E_{X_i} are disjoint and do not interact. Showing just two observers explicitly, the encoding process can be depicted as (cf. [46] Fig. 1):

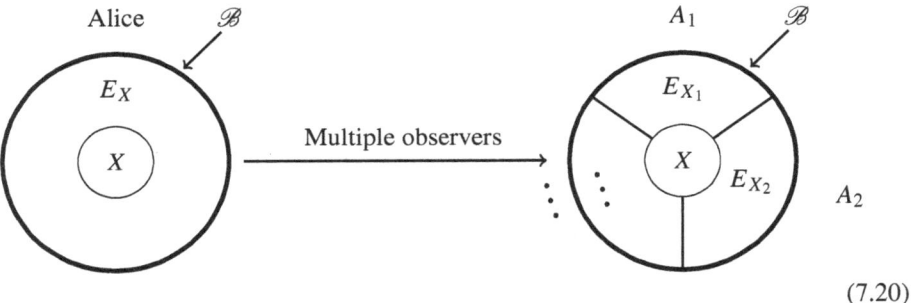

$$(7.20)$$

As formulated in [42, 43], quantum Darwinism assumes that the observers can distinguish, and interact specifically with, the X-specific environmental fragments E_{X_i} without prior communication or prior knowledge of X. This assumption is clearly unrealistic [47]; the observers require X-specific QRFs to distinguish the E_{X_i} from alternative environmental fragments that encode pointer states of other systems. If we add this requirement, then each of the E_{X_i} implements a quantum channel from X to the relevant observer A_i, and hence composite systems of the form $E_{X_i} X E_{X_j}$ implement quantum channels between observers A_i and A_j. If we further allow the observers to agree to observe X and to share their observations via a separate, classical channel, we have a LOCC protocol.

The Quantum Darwinist construction highlights the critical role of redundant encoding of observationally-accessible data on $\mathscr{B}$ for systems A comprising multiple observers. If the observers are to remain distinct, and therefore mutually separable, the results of Sect. 7.2 require that the redundancy of encoding be only approximate, a requirement that is met if the different environmental fractions E_{X_i} impose even slight basis rotations on the data they transport from X. As amply demonstrated in practice, such approximately redundant encodings allow communities of observers, each interacting with the world via their own MB, i.e. their own sector of $\mathscr{B}$ in the classical limit, to correct observational errors. Hence quantum Darwinism is, effectively, a mechanism for constructing an error-correcting code; we will see below that it is in fact a QECC.

Box 7.1: Device-independent methods—physics in a box

Experimentalists have traditionally assumed that they know what system they are making measurements of, and that they know how the experimental apparatus being used to make the measurements works. As pointed out in Chap. 1, these assumptions are challenged whenever either system or apparatus is treated as a black box. While holographic screens or MBs render all systems black boxes as a matter of principle, the assumption that we can "look inside" the box and see how it works can often—and must often, for experimental science to be possible—be made in practice. One situation in which it cannot be made is when "knowledge" of the system can only be obtained indirectly from sources that cannot be trusted. This is the case, for example, in security applications, e.g. quantum cryptography [48], in which any indirect sources of knowledge may in fact be adversaries. *Device-Independent Methods* (DIMs) were developed to maximize the trustworthy information that can be obtained from non-trusted sources; see Grinbaum [49] for a review.

Any quantum process can be "put in a box" that is treated as a black box, i.e. inputs and outputs occur at an interface—an MB—that is taken to be otherwise impenetrable. This is the case, for example, in any scattering experiment—the scattering center is quite literally "in a box" accessed by beam lines and surrounded by detectors. A canonical model for quantum operations is the *(nonlocal) Popescu-Rohrlich (PR) box* [50].

Here is a brief outline of the main ideas of DIMs, using the PR box as an example and following [49]. Consider sets $\mathcal{X}_1, \ldots, \mathcal{X}_n$, and $\mathcal{A}_1, \ldots, \mathcal{A}_n$, to be alphabets of finite cardinality. Consider a set of n parties selecting a measurement setting, and assigning an input value $x_1 \in \mathcal{X}_1, \ldots, x_n \in \mathcal{X}_n$, respectively, then subsequently receiving the results in terms of an output $a_1 \in \mathcal{A}_1, \ldots, a_n \in \mathcal{A}_n$. A conditional probability distribution

$$\mathbf{p} = P(a_1, \ldots, a_n | x_1, \ldots, x_n) \tag{7.21}$$

is hypothesized to encode the 'physics'. In a contextual sense, the distribution $\mathbf{p}$ is *non-signaling*, if and only if all one-party marginal probabilities are functions of their respective inputs x_i: that is:

$$P(a_i | x_1, \ldots, x_n) = P(a_i | x_i) \tag{7.22}$$

In a PR box the unknown processes connect the inputs $x, y \in \{0, 1\}$, and the outputs $a, b \in \{0, 1\}$ of two parties according to the joint distribution:

$$P(ab|xy) = \begin{cases} 1/2 : \ a + b = xy \bmod 2 \\ 0 : \ \text{otherwise.} \end{cases} \qquad (7.23)$$

The idealized concept of the PR box (with an element of the Bayesian principle about it), together with the no-signaling condition, extends quantum theory by violating the Tsirelson bound, but due to the latter condition respects Special Relativity. Further, nothing is assumed about the box's internal functions, and the interior is not specified by any physical theory. A typical DIM, such as a PR box, dispenses with 'globality' between I/O relations, together with the laboratory-environment concept. For the sake of investigating a physical theory, as [49] suggests, the DIMs comprise a "territory of science" that is relatively new to physics with its typically "externalized" observers [51], but that corresponds strikingly to that of the FEP. The DIM as defined by choice of alphabets for the I/O relations, where the former houses the instrumental strings and words, together with the probabilistic conditions, leads to Grinbaum's contention that "physics is about language" [49], which we think of as somewhat complementary to Wheeler's claim that "physics is about information" [14, 52].

7.3   LOCC Protocols Induce QECCs

The idea of a QECC was briefly introduced in Chap. 2. With the above background, it is straightforward to see that any LOCC protocol induces a QECC, or equivalently, that any LOCC protocol executed by Alice requires that Bob implements a QECC (for specific details, see [10, Sect. 4]). Using the notation of Diagram (7.12), A_1 and A_2 can communicate via a LOCC protocol only if Bob's internal interaction H_B implements both quantum and classical channels with sufficient fidelity. We can assume that the classical channel is implemented with sufficient fidelity to allow classical error-correction methods, e.g. parity checking or message repetition; either is just a matter of sufficient bandwidth. The relevant question is assuring the fidelity of the quantum channel. The conditions for doing so are given under generic assumptions by the general theory of QECCs [53], and involve encoding states to be protected into a larger Hilbert space subject to known interactions. These conditions constitute, by definition, those that must generically be met by a QECC. Hence LOCC protocols require QECCs.

As discussed in [53], the task of a QECC is to protect an encoded quantum state from degradation by the environment. In the simplest case, we can regard the encoded quantum state as the state $|S\rangle$ of some sector S of $\mathscr{B}$; in the limit, this state $|S\rangle$ may be the state of a single qubit, as in, e.g. one arm of a Bell/EPR experiment. Treating S as defined by, and hence $|S\rangle$ as prepared by, the "user" of the QECC, Alice, the relevant environment is the entirety of Bob, other than those components that implement the quantum channel and hence the QECC. Adapting the notation of [53] to that of Diagram (7.12), the perturbative action of the environment (i.e. Bob) can be represented by a set of error-inducing operators B_α such that [53, Eq. 5]:

$$\sum_\alpha B_\alpha^\dagger B_\alpha = I. \tag{7.24}$$

Letting 2^k be the dimension of the effective Hilbert space $\mathcal{H}_S$ of the sector S to be preserved, the codespace is a 2^n dimensional Hilbert space $\mathcal{H}_C$, where $n >> k$ and C designates the code. An encoding operator is then a positive, trace-preserving, surjective map $E : S \to C$; a decoding operator is its right inverse D, i.e. $ED = I$. Letting $|0_L\rangle$ and $|1_L\rangle$ designate the logical states of C, C protects against perturbations by the B_α if and only if [53, Eqs. 6, 7]:

$$\langle 0_L | B_\alpha^\dagger B_\beta | 1_L \rangle = 0,$$
$$\langle 0_L | B_\alpha^\dagger B_\beta | 0_L \rangle = \langle 1_L | B_\alpha^\dagger B_\beta | 1_L \rangle. \tag{7.25}$$

As discussed in [53], the B_α are in practice not fully known, so a practical code can only be proved protective against some anticipated subset of the B_α.

The conditions given in Eqs. (7.24) and (7.25) are completely generic, requiring only that the B_α do not act on $|S\rangle$ prior to its encoding by E [53]; see also [54] for further discussion, and in particular [55] which exhibits a means of perfecting QECCs.

Hence we can represent any QECC compactly as a mapping:

$$S \xrightarrow{E} C \xrightarrow{U_C} C \xrightarrow{D} S' \tag{7.26}$$

in which U_C is an automorphism of C (technically, of $\mathcal{H}_C$), S' is the sector onto which the encoded data are decoded, and the composition DU_CE is required to be information-preserving. We can, therefore, represent a QECC generically as a TQFT that implements the quantum channel in a LOCC protocol:

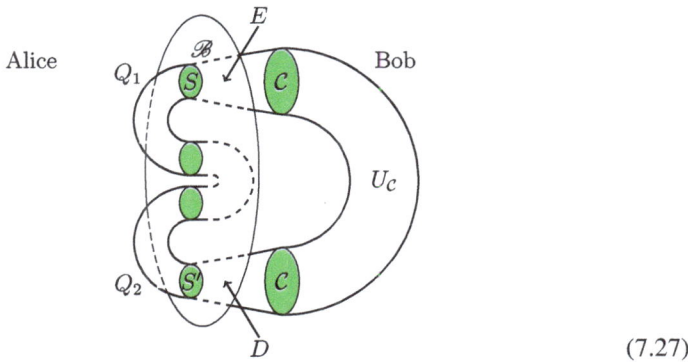

$$\tag{7.27}$$

Some general remarks are now in order:

Remark 7.31 Any QECC requires a classical channel between its users if it is to have practical utility. It is the existence of this classical channel, in particular, that operationally demonstrates the separability of the users. Hence in practice, QECCs are both required by, and require, LOCC protocols.

Remark 7.32 While the above description treats the encoding operator E and its inverse D as implemented by Bob, in practice it is more common to think of these operators as QRFs implemented by Alice. In this case, the codespace $\mathcal{H}_C$ becomes a sector of the boundary $\mathcal{B}'$ between Alice′—i.e. Alice plus the degrees of freedom needed to implement E and D—and Bob′. "Moving the boundary" from $\mathcal{B}$ to $\mathcal{B}'$ is simply moving the "Heisenberg cut" between the observer and the observed system; it changes nothing about the physics. Hence whether we regard the observers or "users" of C as interacting directly with the codespace is a mere notational issue.

Remark 7.33 Consonant with the above, the entanglement that serves as a resource for error correction is always provided by one or both of the two interacting bulk systems, i.e. by either Alice or Bob. The operators E, U_C, and D can all be regarded as QRFs implemented by either Alice or Bob.

Remark 7.34 The "users" of a QECC (here, A_1 and A_2) must be separable both from each other and from whatever bulk system implements the QECC (here, Bob). As discussed above, this is equivalent to noting that classical communication (or classical memory) is a required resource for a useful QECC.

Example 7.3.1 Quantum Darwinism depends on a QECC As discussed in Sect. 7.2.3 above, quantum Darwinism requires that the composite systems $E_{X_i} X E_{X_j}$ between distinct observers A_i and A_j of some quantum system X implement a quantum channel that enables A_i and A_j to each, independently, report an observation of the same state $|X\rangle$ of X. Let us examine this one step at a time. First, we note that the environmental fragments E_{X_i} each provide a continuous path from X to the appropriate A_i in the representation of [45, 46]; hence we can assume that each fragment is simply connected. The choices of the E_{X_i} are arbitrary, and in particular, not optimized in any way by the A_i. Redundant encoding of $|X\rangle$ by the E_{X_i} then requires, as noted earlier, that the local interactions $H_{X E_{X_i}} = H_{X E_{X_j}}$ for all components i, j. Hence the interaction $H_{X E_X}$ between X and its total environment E_X must be uniform, at least at the resolution sampled by the E_{X_i}; it cannot, in particular, depend on any local coordinates that characterize either X or E_X.

Let us now consider the particular fragments E_{X_i} and E_{X_j}, $i \neq j$, and the interactions $H_{X E_{X_i}}$ and $H_{X E_{X_j}}$ defined at the boundaries $\mathcal{B}_i$ and $\mathcal{B}_j$ between E_{X_i} and E_{X_j}, respectively, and X. The requirement that $H_{X E_X}$ be uniform is the requirement that E_{X_i} and E_{X_j} employ the same basis for their local components of the overall interaction $H_{X E_X}$; otherwise the eigenvalues of $H_{X E_X}$ would depend on the fragment chosen. The fragments E_{X_i} and E_{X_j} are, therefore, entangled [5]. It is this entanglement that enforces $H_{X E_{X_i}} = H_{X E_{X_j}}$ and enables—indeed, enforces—redundant encoding of $|X\rangle$ by E_{X_i} and E_{X_j}.

Any state $|X\rangle$ encoded by any of the E_{X_i} must be an eigenstate of the relevant interaction $H_{X E_{X_i}}$ and hence a component of an eigenstate of $H_{X E_X}$; this requirement is termed *eins-election* in [45]. Consider now the consequences of small perturbations of $|X\rangle$ that perturb $H_{X E_X}$ and hence the encoded eigenvalues. We can represent such perturbations as the actions of error-inducing operators B_α that transform, for example, an eigenstate of position into a spatial superposition [42, 43]. Such perturbations are damped out, on average, simply by their stochasticity; in the terminology of [45, 46], this stochastic damping is referred to as the *prediction sieve* implemented by the E_{X_i}. The prediction sieve effectively implements the trace $\mathrm{Tr}_E(\rho_{X E_X})$ over environmental degrees of freedom found in standard environmental-decoherence theory [45]. The encoding is described at equilibrium by a TQFT, while the role of the stochastic perturbations can be included as an out-of-equilibrium QFT-described phase, which finally undergoes relaxation toward the TQFT phase by following a geometric relaxation flow [56].

We can, therefore, see quantum Darwinism as the observation that the environment E_X of any quantum system X is, effectively, a codespace that redundantly encodes eigenvalues of $H_{X E_X}$ and damps out the actions of any "noise" operators B_α that perturb those eigenvalues by perturbing $|X\rangle$. This picture can be effectively extended, so as to account for error-inducing stochastic perturbations acting on the system X, exploiting out-of-equilibrium renormalization group flow dynamics of complex systems [56]. The codespace is "large" in the sense of allowing multiple observers to access the encoded state information without perturbing the encoding. The encoding is uniform, and hence redundant, because $H_{X E_X}$ is uniform. Entanglement between fragments E_{X_i} of E_X—which is undetectable by any observer(s) restricted to one or a few of the E_{X_i}—assures uniformity of $H_{X E_X}$, and hence is the key resource required for redundant encoding.

Example 7.3.2 Bell/EPR experiments depend on a QECC Bell/EPR experiments are some of the most important in contemporary physics [36, 37]. Using the notation above, the system X in a Bell/EPR experiment is an entangled pair, with the state $|X\rangle$ being the singlet $(|\uparrow\downarrow\rangle - |\downarrow\uparrow\rangle)/\sqrt{2}$. There are only two observers, A_1 and A_2. Each observer A_i interacts with a fragment of the environment E_{X_i} comprising a detector, etc. The observers exchange results via a classical channel that enables classical statistical tests, e.g. tests for the violation of Bell's inequality.

A Bell/EPR experiment can, therefore, be seen as an instance of quantum Darwinism that involves two observers interacting with a singlet state. What enables the detection of entanglement, and hence gives the experiment its power, is the arrangement of the environmental fragments E_{X_1} and E_{X_2} in a way that allows each of A_1 and A_2 to access only one component of the singlet; this is accomplished, in practice, by spatially separating the detectors. This restriction of the observers to single state components allows the actions of one observer to influence the outcome observed by other. It effectively converts the QECC described in the example above into an erasure code. To see this, let us examine the experiment in stages. Provided the E_{X_i} are "set" to employ the same z axis, the combined environment E_X damps out the actions of mutually-uncorrelated perturbations B_α of $|X\rangle$ as discussed above. The detectors, in this case, "witness" the arrival of "particles" with classically correlated spins. Allowing the observers to each freely choose a z axis, and hence a measurement basis, between observations introduces a distinct set of perturbations B'_α. While operationally, each of the B'_α perturbs only the local environmental fragment E_{X_i}, Eq. (7.25) requires each of the B'_α to act on the entire state $|X\rangle$. It is this requirement that encodes the effective entanglement[2] between E_{X_1} and E_{X_2}, and hence that a state component "erased" by measurement at E_{X_1} is reproduced, with a spin flip, at E_{X_2}.

We have shown here, mainly via several examples, that TQFTs provide a minimal-assumption language for describing quantum-state preparation and measurement, and hence a minimal-assumption language for describing the multi-agent interactions that implement LOCC protocols. We have also shown LOCC protocols generically induce QECCs, and considered how QECCs are induced in multi-observer settings described by quantum Darwinism and in two-observer Bell/EPR experiments. These examples both illustrate the dependence of a QECC on a uniform choice of basis, and hence the dependence of a QECC on LOCC. We can, therefore, see LOCC protocols and QECCs as generically equivalent in practice: a LOCC protocol is useless unless the quantum channel is robust against noise and hence functions as a QECC, while a QECC is useless to observers who have no means of agreeing what basis to employ.

It is interesting to note that in both quantum Darwinism and Bell/EPR experiments, the bulk entanglement induces, via the action of a QECC, redundant classical encodings at spatially-separated locations, i.e. at the locations of spatially-separated observers. As pointed out by Einstein [58], spatial separation is a resource for separability, and hence for effective classicality. The generality of the methods employed in Sect. 7.3 suggests that this outcome is completely general: that bulk entanglement generically induces classical redundancy via spatial separation. The task of showing that this is the case—that spacetimes can, generically, be regarded as QECCs, is addressed in the next section. This result yields an intriguing hypothesis, one consistent with remarks already made in [12]: that the fundamental role of spacetime in physics is to provide "a place to put" redundancy. Indeed it suggests that the concept of redundancy as a necessary resource for communication and the concept of spacetime as a necessary resource for dynamics may, at some fundamental level, be the same concept. This being the case, it would substantially support Wheeler's contention [14] that physics is fundamentally about information exchange. The overall train of ideas further suggests support for the suggestion of Grinbaum [49] that physics is also fundamentally about language, as highlighted previously in the context of DIMs.

---

[2] Note that Entanglement Purification Protocols on certain mixed states sharing halves of EPR pairs through a channel, can also yield a QECC, and conversely [57].

7.4   Spacetime as a QECC

Following the groundbreaking work of Peter Shor on polynomial-time factorizing algorithms using powerful resources of quantum computation and, in particular, implementing QECCs to deal with frequent error-prone qubits [59, 60], Almheiri, Dong and Harlow [61], in view of the AdS/CFT correspondence [62], posited spacetime as a QECC (reviewed in [63]). Here, we shall enhance matters by first taking another look at Diagram (7.27). We recall that both the sectors S and S' of the A-B boundary $\mathscr{B}$, and the codespace $\mathcal{C}$ can, without loss of generality, be represented as finite qubit arrays [4, 5]. As we have pointed out, the sectors S and S' must be separable if they are to correspond to observers able to make independent observations. The quantum channel $EU_{\mathcal{C}}D$ increases (via the encoding E) and then decreases (via the decoding D) the dimensionality with which the state $|q_i\rangle$ of any qubit q_i in the boundary sector S is represented by qubits of the bulk, effectively entangling q_i with some qubit q_i' in S'. Classical data encoded by A_1 on S_1 can, therefore, be recovered, up to perturbations imposed by B, by A_2 from S_2. This classical redundancy between sectors is enabled by the requirement that A_1 and A_2 be mutually separable, as discussed earlier. A Bell/EPR experiment provides an example of how the encoding of classical data may be implemented by instrument settings that are, effectively, communicated to the other observer as observable perturbations of the shared quantum state $|X\rangle$.

Now, it is reasonable to think that nothing prevents the extension of quantum Darwinism to arbitrary, finite numbers of observers; indeed this is contemplated already in [42, 43]. Hence we can regard the boundary $\mathscr{B}$ as a manifold arbitrarily tessellated, and regard each tile as a sector with an associated observer. In this case, separability of $\mathscr{B}$ as a manifold corresponds to separability of the associated observers, and allows the observers to each, independently, encode or report classical data. The bulk B supports error correction for these classical data—and hence enables classical redundancy—if it implements a codespace $\mathcal{C}$ that the ith observer can access via encoding and decoding operators E_i and D_i as depicted in Diagram (7.27). Hence the possibility of redundant encoding on all of $\mathscr{B}$ depends on B implementing a single QECC equally accessible from every boundary sector. This condition is implemented precisely by *bulk-boundary codes* such as those induced by the AdS/CFT correspondence [62].

7.4.1   Example: Bulk-Boundary Codes, Entanglement Wedges, and Scrambling

We can further explore matters in terms of this example [64]. With this in mind, let us return again to Diagram (7.27) where two obvious questions turn up. Firstly, how large does $\mathcal{C}$ have to be to protect the data encoded on S, and secondly, what is the relationship between the time evolution $U_{\mathcal{C}}$ of the codespace, and that of the bulk B that implements it? The latter can be expressed, in the language of [53], as the question of what perturbations B_{α} we can expect B to impose on C. These questions are addressed in [61] in the context of AdS/CFT, assuming that the B_α are strong enough to erase some of the information in C; that is,

$$B_\alpha : |\uparrow\rangle, |\downarrow\rangle \mapsto |0\rangle =_{Def} (|\uparrow\rangle \pm |\downarrow\rangle)/\sqrt{2} \qquad (7.28)$$

for some subset e of the qubits in C. There is also the question of how large C must be to correct for—or be impervious to—such erasures. If k qubits are required to encode a message without error correction, and l qubits are erased, C requires [61, Eq. 3.22]:

$$n \geq 2l + k \qquad (7.29)$$

qubits for error correction. As remarked in [61], this is an intuitive result: for each qubit that is erased, at least two others are required to reconstruct its state. Rather than the size of a codespace suited to error correction, it is its entanglement structure that is of significance to us.

As further demonstrated in [61], the codespace C can be protected against the erasure of a set e of qubits only if [61, Eq. 3.14]:

$$\rho_{er}[\phi] = \rho_e[\phi]\rho_r[\phi] \qquad (7.30)$$

where r is the set of k qubits added to the codespace to enable protection and all partial traces are over the entire n-qubit state ϕ. Separability between qubits in e and those in r clearly depends on the choice of basis for ϕ [31]. Again, there is dependence of the QECC on the availability of a classical channel facilitating use of the code to coordinate choice of the same basis for the users local E and D operators; i.e. the dependence of the QECC on LOCC at the level of the observers/users. Quantum redundancy within the QECC thus depends on classical redundancy of basis choice; that is, on gauge invariance [12].

We are also interested in the extent to which a qubit q in C is protected from erasure functions, in an AdS/CFT context, as an effective radial coordinate for the AdS bulk—qubits in the "center" of the bulk are easily protected, while qubits near the boundary are more vulnerable (see [61]). Referring again to the generic representation of Diagram (7.27), we note that the entire AdS bulk is treated as the codespace in [61], and that both the time evolution U_C, and the encoding and decoding operators E and D are left implicit. Let us consider the boundary CFT to be defined on $\mathscr{B}$, and consider each neighborhood of $\mathscr{B}$ to be a sector S_i characterized by a local CFT observable x_i. Decoding and encoding the state $|x_i\rangle$ on S_i correspond, respectively, to obtaining a classical outcome value by measuring $|x_i\rangle$ using a local QRF Q_i and to preparing $|x_i\rangle$, using Q_i, given such a classical value; measuring and preparing the state $|\uparrow\rangle$ of a single qubit provides an example. The bulk neighborhood C_i of S_i on which $|x_i\rangle$ is encoded is, as pointed out in [61], just the causal wedge $\mathscr{W}_C[S_i]$. Encoding $|x_i\rangle = |0\rangle$ on S_i erases the qubits in $\mathscr{W}_C[S_i]$, but has negligible effect in the rest of C, which encodes the combined values of $|x_j\rangle$ on S_j for $j \neq i$. We can, indeed, think of the qubits in $\mathscr{W}_C[S_i]$ as implementing local operators E_i and D_i that act on the rest of C, i.e.

on $\mathcal{C} \setminus \mathscr{W}_{\mathcal{C}}[S_i]$. In the language of quantum Darwinism, the $\mathscr{W}_{\mathcal{C}}[S_i]$ are the 'environmental fragments' with which the local (to S_i) observers interact.

Harlow [65] has further explored this connection between proximity to the boundary and vulnerability to erasure by establishing an equivalence relation between erasure protection and the Ryu-Takayanagi generalization [66] of the Hawking-Bekenstein black hole entropy. If X is a d-dimensional system, with a $(d-1)$-dimensional boundary ∂X in a $(d+1)$-dimensional CFT, and γ_X is a d-dimensional surface in $\mathrm{AdS}_{(d+2)}$ with boundary ∂X, then following [66, Eq. 1.5], the entanglement entropy is defined by:

$$S_X = \frac{\text{Area of } \gamma_X}{4 G_N^{(d+2)}} \tag{7.31}$$

where $G^{(d+2)}$ is the gravitational constant in the bulk $\mathrm{AdS}_{(d+2)}$. If Ξ_X is the bulk region bounded by $\gamma_X \sqcup X$, the entanglement wedge is recovered:

$$W[X] = D_{\text{bulk}}[\Xi_X] \tag{7.32}$$

where D_{bulk} is the bulk domain of dependence. We can, therefore, interpret Eq. (7.31) as relating the entanglement entropy of a system X in the boundary CFT to the entanglement entropy of its corresponding wedge $W[X]$ in the bulk [63]. Erasing quantum information in one system, therefore, erases it in the other. Related is the approach adopted in [67], in which redundancy and hence error-correction capacity on the boundary is again provided by entanglement within the bulk. Scrambling processes effectively remove redundancy from the bulk, and hence from any QECC implemented by the bulk [68, 69]. Thus scrambling in the bulk generates noise on the boundary.

One of the main features of [64] was to show that, given appropriate symmetries on $\mathscr{B}$, a TQFT on $\mathscr{B}$ can be represented by a *topological quantum neural network* (TQNN)[3] on $\mathscr{B}$ that effectively encodes topological connectivity as metric connectivity (see e.g. [71]). The TQFT method and generalizations for *Deep Neural Networks* (DNN) are particularly important [72, 73]; in fact, a DNN can be considered as the semi-classical limit of a generalized QNN to a TQNN [74]. A number of associations of TQNNs in relationship to QECC and various spacetime models (e.g. Chern-Simons, BF and Kähler space-time) are discussed in [64]. It is in this context that we can discuss the 'perfect tensors' and the 'HaPPY' code of [75] where perfect tensors represent erasure-protection, optimally efficient QECCs.[4]

---

[3] Essentially, this consists of the dynamics induced by states of the spin-network basis on a Hilbert space of the TQFT [70].

[4] A *perfect tensor* is a tensor T with m indices, such that for any bipartition of the indices into sets of n_1 indices and n_2 indices, $n_1 + n_2 = m$, T is a proportional to an isometric tensor from V_1 to V_2, where V_i is the vector space spanned by the n_i indices of T. As pointed out in [75], T is perfect if T is unitary and $n_1 = n_2 = n$. In this case, the tensor describes a pure state of $2n$ spins such that any set of n spins is maximally entangled with the complementary set of n spins. It is this maximal entanglement condition that renders the 'HaPPY' code constructed with such a tensor optimally efficient for erasure protection.

7.5   ER = EPR is an Operational Theorem

We recall that a conjecture of Maldecena and Susskind [76] in the framework of AdS/CFT duality, states that an *Einstein-Rosen (ER) bridge* (sometime called a 'wormwhole') is equivalent to a pair of maximally *Einstein-Podolski-Rosen* (EPR) entangled black holes; hence that ER=EPR (this is also discussed in [77]). Whereas the approach of [76] is essentially geometric, we generalized matters by topological methods in [78] in showing that two distinct observers cannot operationally distinguish, by independent local manipulations and measurements, monogamous entanglement from a topological identification of points in their respective local spacetimes. This means that in a two-agent setting, it is not possible to *operationally* distinguish ER from EPR. The method of achieving this involves LOCC protocols as above; we shall proceed with outlining the steps following [78].

7.5.1   A Canonical Bell/EPR Experiment

Firstly, we can adopt the setting of a canonical Bell/EPR experiment. Given this setting, Alice (A) and Bob (B) can only detect entanglement if they each have a free choice of measurement basis; that is, only if each of them can independently choose the instrument settings to employ for each measurement. If this free choice requirement is violated by some form of superdeterminism (or a "conspiracy"), A and B are themselves effectively entangled, and hence cannot function as two independent observers. To talk about "classical communication", if Alice and Bob cannot be considered separate systems, is patently spurious. Hence formally, separability of the joint state, $|AB\rangle = |A\rangle|B\rangle$ (or of the joint density $\rho_{AB} = \rho_A\rho_B$) is a requirement of LOCC, and hence of operational access to a shared quantum channel [10]. How these requirements unfold can be summarized as follows:

(1) Alice and Bob both perform only *local* operations, and therefore, each must employ spatial QRFs [8]. These we will denote by X_A and X_B, respectively, with respect to which they specify the position of the quantum degrees of freedom that they manipulate, that is, the positions of the detectors in a Bell/EPR experiment. These spatial QRFs must commute with the QRFs Q_A and Q_B that they, respectively, employ to manipulate the quantum channel, that is, $[X_A, Q_A] = [X_B, Q_B] =_{def} 0$.

(2) Alice and Bob must both comprise sufficient degrees of freedom for the purpose of i) implementing their respective QRFs, and ii) communicating classically. This is amounts to a large N-limit that assures their separability as physical systems.

In Diagram (7.12) above, both classical and quantum channels are implemented by E. Hence, they are in general, exposed to interaction with other qubits contained within E. In a descriptive sense, Alice and Bob can be seen to be:

(1) connected by a quantum channel if there exist distinct (collections of) qubits q_A and q_B accessible only to Alice and Bob, respectively, and $|q_A q_B\rangle \neq |q_A\rangle |q_B\rangle$;

(2) connected by a classical channel if there exist distinct (collections of) qubits q'_A and q'_B accessible only to Alice and Bob, respectively, and causal processes f and g implemented by E, such that $q'_B = f(q'_A)$, and $q'_A = g(q'_B)$.

The sense of 'causal' is in relationship to clocks in E. Alice and Bob being distinct, mutually separable agents that communicate classically (that is, causally) implies that the QRFs Q_A and Q_B do not commute, and hence that LOCC protocols exhibit quantum contextuality in the sense we had discussed in Chap. 5. We consider Alice and Bob to also be separated by a boundary $\mathcal{B}$, as in (7.12), to emphasize their joint interaction with E, and to have access only to distinct, non-overlapping sectors of $\mathcal{B}$.

Both the classical and quantum channels in (7.12) are implemented by E, and hence are, in general, exposed to interaction with other qubits contained within E. Such interactions between the degrees of freedom of E that implement the channels, and other, non-channel degrees of freedom of E, can induce eavesdropping, or, generate noise into the classical channel, and decoherence into the quantum channel. In both cases, the coupling of the channel to the non-channel degrees of freedom is a quantum interaction, the result of which is to transform pure states of channel degrees of freedom into mixed states. When such mixing happenings, observations of the channel states, whether classical or quantum, will be characterized by *stochastic noise*, with a uniform noise spectra in the limit of maximally mixed states. The situation is that Alice and Bob have no direct observational access to such mixing interactions, which occur entirely within E, and are implemented by the internal interaction Hamiltonian H_E. The risk of decoherence can be minimized by minimizing the exposure of the quantum channel to the rest of E. Formally, such procedures correspond not to tracing out degrees of freedom of E, but to setting their interaction with the channel degrees of freedom to zero.

7.5.2   The Main Result

Let us now consider a quantum channel in the limit of zero decoherence. In an idealized Bell/EPR experiment, this corresponds to a 'monogamous' pairwise entanglement. Technically, this results in a Bell inequality violation reaching the Tsirelson bound, provided Alice and Bob consistently choose settings 45° apart. Setting decoherence equal to zero is requiring that there no interaction between channel and non-channel degrees of freedom of E. Hence the limit of zero decoherence in the quantum channel is attained as the interaction in E, between channel and non-channel degrees of freedom, approaches zero. Taking q_A and q_B to be the (collections of) qubits accessible exclusively to Alice and Bob, respectively as above, the limit of zero decoherence is reached when the state of the quantum channel is simply the pure state $|q_A q_B\rangle$.

We can assume, without loss of generality, that Alice and Bob interact directly with the qubits q_A and q_B to which they have their exclusive access. We can, therefore, assume that these qubits are localized to Alice's and Bob's respective sectors of the boundary $\mathscr{B}$. The pure state $|q_A q_B\rangle$ is, in this case, an entangled state of $\mathscr{B}$. The following diagram [78, (2)] illustrates how all of this is configured:

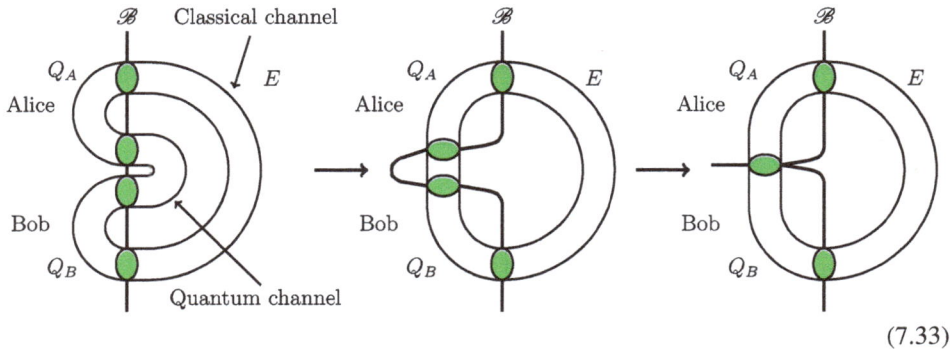

$$(7.33)$$

Diagram (7.33) is amenable to topological transformations, as would be the case in a Bell/EPR experiment, where such a transformation would be approximated by decreasing the laboratory-frame distance between each of Alice and Bob, and the centrally-located source of entangled pairs toward zero. In which case, the limit shown in the rightmost diagram represents a point source to which Alice and Bob are both immediately adjacent. As discussed in [6], it also corresponds to a Bell/EPR experiment as described from the perspective of the entangled state $|q_A q_B\rangle$, in which the observers Alice and Bob effectively collide at the fixed position of $|q_A q_B\rangle$ (see [6, Fig. 8a]).

Returning to the context of LOCC protocols, the main result of [78] that leads to showing that ER=EPR is an operational theorem, is:

Theorem 7.5.1 *In any LOCC protocol in which all systems are finite, and in which the boundary $\mathscr{B}$ between the communicating agents A and B and their joint environment E is a holographic screen, as the entanglement made available to A and B by the quantum channel approaches pairwise monogamy, and hence the decoherence in the quantum channel detectable by A or B decreases to zero, the number of environmental degrees of freedom of E required to implement the quantum channel becomes operationally indistinguishable, by A or B, from zero in the limit of monogamous entanglement.*

There are two significant consequences of Theorem 7.5.1:

(1) *The codespace dimension of a perfect QECC is operationally indistinguishable from the code dimension* [78, Corollary 1].
(2) In specifying the connections with ER and EPR, we have: *In any LOCC protocol in which all systems are finite, and in which the boundary $\mathscr{B}$ between the communicating agents A*

and B and their joint environment E is a holographic screen, a quantum channel imple-
menting a shared, monogamously-entangled pair of qubits ("EPR") is operationally
indistinguishable from a topological identification of the locally-measured locations x_A
and x_B *of the qubits accessed by A and B, respectively ("ER")* [78, Corollary 2].

Note that since the last result does not assume an embedding geometry, but only assumes
quantum and classical channels in a topological framework between Alice and Bob, namely
LOCC, it substantially generalizes both the original formulation of [76], and subsequent
geometric formulations of ER = EPR (such as relative to the geometry of ER-bridges, as
described in [79], for instance).

7.6   Nonlocal Resources for Quantum Computation

Implementing both spatial and temporal resources within computational processes overall
determine what we may call *computational complexity*. The resource demands of these
processes is a question of deep interest which we will approach here. In particular, we will
address those computational processes that utilize nonlocal spatial, or temporal resources as
these pertain to computation in general relativity and quantum theory. These are questions
closely tied to the infamous P = NP problem. Specifically, as studied within the above
framework of LOCCs and ER = EPR as an operational theorem, we will further bring to the
table:

(1) *Nonlocal, multiprover interactive games* (e.g. [80]).
(2) *Closed timelike curves* (e.g. [81]).
(3) *Constraint satisfaction problems* (e.g. [82]).

These, we will view in the light of TM-undecidable problems such as the Halting problem
with respect to the class of problems of *recursively enumerable languages* (RE) equivalent
to the Halting problem (for an explanation, see Appendix E on Rice's Theorem).

7.6.1   Nonlocal Games

A *nonlocal game* is a keystone concept in quantum/computation/information. Basically, such
a game proceeds by the interaction of three parties. These consist of two noncommuting
players or *provers* A and B, and *verifier* or *referee* C. Players A and B are allowed to
communicate on a classical level before the start of play, but not thereafter; they can also
share an arbitrary, bipartite state. The verifier C samples a pair of questions drawn from some
distribution, and then separately sends one of them to each of A and B. It is pertinent that
each of A and B answers *classically* to the verifier. They win the game when the questions

and answers satisfy a given predicate, given A and B know the distribution of questions and the predicate. A *quantum value* can be assigned as the supremum of the probability that A and B win the game. All of this can be generalized to fully nonlocal quantum games in the possible presence of noise [80, 83, 84].

The above outline of a two-prover nonlocal game can be extended to multiple provers. A *multiple interactive prover* (MIP) game, as was introduced in [85], now consists of multiple provers who, likewise, can communicate with each other prior to posing the problem, but not thereafter. The idea of the game, spanning k rounds with p players, is that the latter contend to convince a polynomial time verifier that a string x belongs to some language $\mathcal{L}$. This nonlocal game is denoted MIP(p, k) in which it is seen that two players are always sufficient for determining the outcome; thus, MIP(2, k) is usually studied (and often with $k = 1$) [86, 87].

If now shared entanglement allowed, then we arrive the class MIP*. It is a spectacular result of [80] that MIP*=RE. In terms of quantum values, as noted in [83], a consequence of this result is that approximating the quantum value of a fully nonlocal game, is undecidable (as mandated by RE). The crucial point here, is that the operational configuration employed in [80] in defining an MIP* machine is a LOCC protocol. What this means is that two independent, and hence separable, provers A and B communicate classically via a TM verifier C who poses the questions and checks the answers while sharing an entangled pair as a quantum communication channel Q. It is often sufficient to check the main result for the case MIP*(2, 1).

7.6.2   Why MIP* Machines Are Not Operationally Identifiable

The aim here is to apply Theorem 7.5.1 to the operational context of a verifier C interacting with an MIP* machine, we will supplement diagram (7.12) by including C as depicted below:

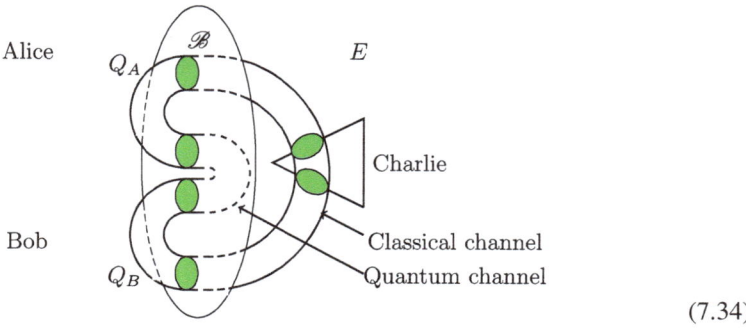

$$(7.34)$$

Here C (Charlie) interacts with A (Alice) and B (Bob) separately, and only via a classical channel as the definition of MIP* necessitates. The next step is to recall, and then put into play Theorem 7.5.1 above, as was the case in [88] in which a technical argument establishes [88, Lemma 1]:

Lemma 7.6.1 *C cannot distinguish data A_i, B_i sent by A and B via a classical channel from measurements of E_C using observables $\hat{A}_i$, $\hat{B}_i$ that yield outcomes A_i, B_i.*

It is important to note that all instances of classical communication require a quantum measurement, by the receiving system of some physical encoding of the communicated information. This particular observation appears to be generally accepted and has been emphasized by Tipler [29]. The reasoning employed in Theorem 7.5.1 and Lemma 7.6.1 now leads to [88, Theorem 2].

Theorem 7.6.1 *An observer C embedded in an environment E cannot determine, either by monitoring classical communication between A and B, or by performing local measurements within E, whether or not A and B are employing a LOCC protocol with classical and quantum channels traversing E.*

Some remarks are in order here. Firstly, the above situation falls into the case of the well-known "conspiracy" or "superdeterminism" theories [89], because C is unable to distinguish, on the basis of reported observational outcomes, between A and B sharing an entangled state and A and B being components of an entangled state. Also, by treating A and B as "effectively classical" experimenters jointly manipulating an entangled state, while remaining separable from each other, as mandated by the definition of LOCC, therefore amounts to a "for all practical purposes (FAPP)" [90] assumption, not a demonstrable fact ([49] provides a general discussion of such circumstances).

Secondly, it is also the case that no component $C \subseteq E$ can determine the entanglement entropy of A and B [88, Lemma 2], which implies that C cannot determine, either by monitoring classical communication between A and B, or by performing local measurements, that A and B are separable (cf. [29]). Consequently, C cannot operationally distinguish between a MIP* machine and some monolithic quantum computer. Any claim that a MIP* machine has solved a TM-undecidable problem is, therefore, operationally circular, as the problem of deciding whether a physical system is actually a MIP* machine is itself TM-undecidable.

7.7 Closed Timelike Curves

Closed timelike curves (CTCs) have been studied from aspects of both cosmology and computation (for a historical survey, see e.g. [91]). For intstance, Hawking, Thorne, and others [92, 93] investigated CTCs mainly in the context of Black Hole theory (e.g. the construction and stability of ER-bridges). When introduced into models of classical computation, CTCs make it possible to solve hard computational problems in constant time (surveyed in [94]). Much can be attributed to Deutsch in [81] who demonstrated that quantum computation with quantum data capable of traversing CTCs, provided a new and powerful physical model of computation, along with self-consistent evolution further engendering (quantum) computational complexity [95]. The formulation of [81] was seen as a basis for a Universal Quantum TM.

As pointed out in [96], one of Deutsch's ideas was to treat a CTC as a region of space-time where a 'causal consistency' condition is imposed; specifically, a region in which the time-evolution operator maps state of the initial hypersurface to itself. This initial state is, therefore, a probabilistic fixed point of the time-evolution operator within the CTC, i.e. a state ρ such that $\Phi(\rho) = \rho$ for the time-evolution operator Φ within the CTC.

Let us give a brief outline of how then computation is modeled by implementing a CTC. We commence with a Hilbert space of qubits given by $\mathcal{H} = \mathcal{H}_{\text{ch}} \otimes \mathcal{H}_{\text{tv}}$, where $\mathcal{H}_{\text{ch}}$ denotes that of the chronologically respecting qubits, and $\mathcal{H}_{\text{tv}}$ that of those which traverse CTCs, as shown in Diagram (7.35) below (cf. [81, Fig. 3] and [95, Fig. 4]):

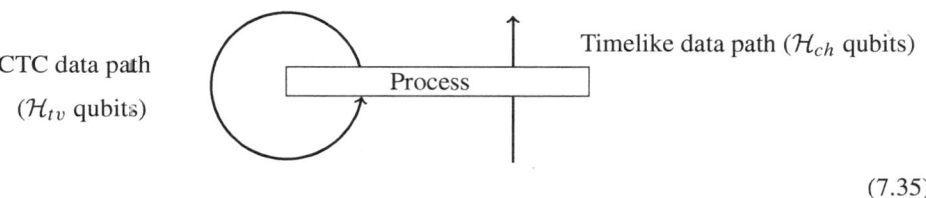

CTC data path

($\mathcal{H}_{tv}$ qubits)

Process

Timelike data path ($\mathcal{H}_{ch}$ qubits)

$$(7.35)$$

It is important to mention that the evolution of CTC qubits is determined by self-consistency—though the qubits themselves are an expendable resource [95]. This means that the state of the CTC qubits, at the temporal origin, should be the same as those qubits after the evolution $\mathbf{U}$ operator corresponding to 'Process' as depicted in (7.35). The density matrix ρ as a solution at the former, is given by

$$\rho = \text{Tr}_{\text{ch}}[\mathbf{U}(\rho_{\text{in}} \otimes \rho)\mathbf{U}^\dagger] \tag{7.36}$$

where ρ_{in} denotes the density matrix of the chronologically respecting qubits, and Tr_{ch} denotes the trace of $\mathcal{H}_{\text{ch}}$ [81, 95]. Given the presence of a quantum circuit, and the solution in (7.36), the output ρ_{out} of the circuit is given by [81, 95]:

$$\rho_{\text{out}} = \text{Tr}_{\text{tv}}[\mathbf{U}(\rho_{\text{in}} \otimes \rho)\mathbf{U}^\dagger] \tag{7.37}$$

There are number of technical issues at stake here for which we refer the interested reader to [81, 95]. For instance, (7.36) and (7.37) assume a 'gating-free' system. When gating is applied, then the consistency condition changes, leading to a previously selected temporal origin as now becoming arbitrary. It is shown, however, in [95] that potentially different self-consistency solutions are relatable via a standard change of basis.

We now approach the question of non-operational identification as that applies to CTCs. Let us then look back at Diagram (7.34). We can regard the joint system AB as some quantum computer, and by setting the decohering interaction H_{CE} to zero, or equivalently, by using the result of Theorem 7.5.1, we treat Q as an internal quantum resource used by AB. The question then arises: what can C infer about the computational role of the classical channel connecting "components" A and B? Since this channel is classical, it requires, by definition, that it is timelike as measured by clocks in E. Taking the limit as $C \rightarrow E$, the channel is timelike as measured by C's clocks. The classical nature assigned to C also requires that the channel has finite length; specifically, the endpoints of the channel, which we can denote A_c and B_c respectively, must be such that $d_C(A_c, B_c) > 0$ in C's distance metric d_C. But it is the case, following [88, Lemma 2], that C cannot determine the entanglement entropy of any state $|AB\rangle$. Hence C cannot determine that A and B are separable as discussed above. We further specify this by means of the following [88, Lemma 3].

Lemma 7.7.1 *In any physical setting described by Diagram* (7.34), *C cannot determine the distance* $d_{AB}(A_c, B_c)$, *where* d_{AB} *is the metric employed by* AB, *between the classical channel endpoints* A_c *and* B_c *on* $\mathscr{B}$.

Following this, it is deduced from [88, Theorem 3].

Theorem 7.7.1 *In any physical setting described by Diagram* (7.34), *C cannot determine whether* AB *employs CTCs as computational resources.*

Of the possible consequences to Theorem 7.7.1, there is one worth mentioning here. It is shown in [96] that both classical and quantum computers can utilize CTCs to solve any problem in PSPACE; namely the complexity class consisting of all problems solvable on a classical TM possessing a polynomial amount of memory. Theorem 7.7.1 shows that the problem of deciding whether a physical system is a computer employing the resource of CTCs, is TM-undecidable. It follows that a proffered solution to a PSPACE problem, for which independent means of verification are unavailable, is also TM-undecidable.

7.7.1   Constraint Satisfaction Problems

Taking stock of what we have shown above, a verifier C cannot operationally distinguish a MIP* machine from a quantum TM, and hence cannot operationally demonstrate the solution to some problem classified in RE. Let us now turn to *Constraint Satisfaction Problems* (CSPs) [82]. In [82, 97] it is shown that CSPs can be formulated in the languages of MIP and MIP* machines, with C administering the satisfaction condition(s). Specifically, [82, Sect. 4] and

[97, Theorem 1.1] demonstrate the relations between CSPs, languages in MIP* (and hence in RE), and protocols for the Halting problem of the form CS-MIP*(2, 1, c, s), with c and s being the completeness and soundness probabilities, respectively (see [82, Corollary 4] for the special case where $c = 1$). The results of Sect. 7.6.2 as applied to MIP and MIP* show that C cannot operationally demonstrate independence between constraints, and identified partial solutions; hence, this also applies to protocols of the form CS-MIP*(2, 1, c, s) as special cases.

In summary, the results of Sect. 7.6 lead to a straightforward interpretation. Namely, finite interactions with an unknown quantum system can place a lower limit, but not an upper limit, on the Hilbert-space dimension of that system. This extends to quantum systems as a limitation on inferences from finite observations that are proved for classical systems (see [98]). The existence of such limits illustrates the profound distinction between behaviors that can be shown theoretically to be logically possible and behaviors that can be unambiguously observed by finite agents such as ourselves. Finally, we recall that the Halting problem has been shown to be equivalent to the Frame Problem [99]. As discussed in Box 5.2, the latter is, broadly speaking, the problem of circumscribing what is relevant in a given situation. This problem can seen to be tied to our discussion as it unfolded in this section; namely, a problem of empirically determining resource availability and its usage.

References

1. Chitambar E. Leung, D., Mančinska, L., Ozols. M. and Winter, A. Everything you always wanted to know about LOCC (but were afraid to ask). Commun. Math. Phys. 2014; 328: 303–326.
2. Atiyah, M. F. Topological quantum field theory. *Publ. Math. IHÉS* 68 (1988), 175–186.
3. Quinn, F. *Lectures on axiomatic topological quantum field theory*. In Geometry and quantum field theory (Lecture notes from the graduate summer school program, June 22–July 20, Park City UT, USA) Amer. Math. Soc., Providence RI. Institute for Advanced Studies, Princeton NJ (1995), pp. 323–453.
4. Fields, C. Glazebrook, J. F.; Marcianò, A. The physical meaning of the Holographic Principle. Quanta 2022; 11:72–96.
5. Fields, C., Glazebrook, J. F. and Marcianò, A. Sequential measurements, topological quantum field theories, and topological quantum neural networks. Fortschritte der Physik 2022; 70: 202200104.
6. Fields, C.; Glazebrook, J.F.; Marcianò, A. Reference frame induced symmetry breaking on holographic screens. Symmetry 2021, 13, 408.
7. Fields, C.; Friston, K.J.; Glazebrook J.F.; Levin M. A free energy principle for generic quantum systems. Prog. Biophys. Mol. Biol. 2022, 173, 36–59.
8. Bartlett, S.D.; Rudolph, T.; Spekkens, R.W. Reference frames, superselection rules, and quantum information. Rev. Mod. Phys. 2007, 79: 555–609.
9. Wootters, W. T.; Zurek, W. H. A single quantum cannot be cloned, Nature 1982; 299: 802–803.
10. Fields, C., Glazebrook, J. F. and Marcianò, A. Communication protocols and QECCs from the perspective of TQFT, Part I: Constructing LOCC protocols and QECCs from TQFTs. Fortschritte der Physik 2024; 72: 202400049
11. Fields, C.; Glazebrook, J.F. Representing measurement as a thermodynamic symmetry breaking. Symmetry 2020, 12, 810.

12. Addazi, A.; Chen, P.; Fabrocini, F.; Fields, C.; Greco, E.; Lutti, M.; Marcianò, A.; Pasechnik, R. Generalized holographic principle, gauge invariance and the emergence of gravity à la Wilczek. Front. Astron. Space Sci. 2021; 8: 563450.

13. Fields, C. and Glazebrook, J. F. (2019a). A mosaic of Chu spaces and Channel Theory I: Category-theoretic concepts and tools. *Journal of Experimental and Theoretical Artificial Intelligence* 31(2), 177–213.

14. Wheeler, J. A. Law without law. In (Wheeler, J.A. and Zurek, W.H. eds.) *Quantum Theory and Measurement* pp. 182–213. Princeton University Press, Princeton, NJ, USA, (1983)

15. Pearl, J. Probabilistic Reasoning in Intelligent Systems: Networks of Plausible Inference; Morgan Kaufmann: San Mateo, CA, USA, 1988.

16. Fields, C.; Glazebrook, J. F. Separability, contextuality, and the quantum Frame Problem. International Journal of Theoretical Physics 2023; 62: 159.

17. Mermin, D. Making better sense of quantum mechanics. Reports on Progress in Physics 2017; 82(1), 012002.

18. Fuchs, C. QBism, the perimeter of Quantum Bayesianism. arxiv:1003.5209v1, 2010.

19. Bohr, N. *Atomic Theory and the Description of Nature.* Cambridge University Press, Cambridge, UK, (1934).

20. Everett, H. III. Relative state formulation of quantum mechanics. *Rev. Mod. Phys.* 29 (1957), 454–462.

21. Rovelli, C. Relational quantum mechanics. *Int. J. Theor. Phys.* 35 (1996), 1637–1678.

22. Mermin, N. D. What is quantum mechanics trying to tell us? *Am. J. Phys.* 66 (1998), 753–767.

23. Tegmark, M. The interpretation of quantum mechanics: Many worlds or many words? *Fortschr. Phys.* 46 (1998), 855–862.

24. Fields, C.; Glazebrook, J. F. Information flow in context-dependent hierarchical Bayesian inference. *J. Expt. Theor. Artif. Intell.* 34 2022, 111–142.

25. Dzhafarov, E. N. Cervantes, V. H. and Kujala, J. V. (2017b). Contextuality in canonical systems of random variables.*Philosophical Transactions of The Royal Society A* 375, 20160389

26. Abramsky, S., Brandenburger, A.: The sheaf-theoretic structure of non-locality and contextuality. New J. Phys. 13, 113036 (2011).

27. Adlam, E. 2021 Contextuality, fine-tuning and teleological explanation. *Found. Phys.* 51 2021, 106 https://doi.org/10.1007/s10701-021-00516-y

28. Abramsky, S.: Contextuality: At the Borders of Paradox. In E. Landry, ed., Categories for the Working Philosopher, Oxford (2017) online edn, Oxford Academic, https://doi.org/10.1093/oso/9780198748991.003.0011

29. Tipler, F. Quantum nonlocality does not exist. *Proc. Natl. Acad. Sci. USA* 111 (2014), 11281–11286.

30. Wootters, W. K. Entanglement of formation and concurrence. *Quant. Inform. Comp.* 1 (2001), 27–44.

31. Zanardi, P. Virtual quantum subsystems. *Phys. Rev. Lett.* 87 (2001), 077901.

32. Zanardi, P., Lida, D. A. and Lloyd,S. Quantum tensor product structures are observable-induced. *Phys. Rev. Lett.* 92 (2004), 060402.

33. de la Torre, A. C., Goyeneche, D. and Leitao, L. Entanglement for all quantum states. *Eur. J. Phys.* 31 (2010), 325.

34. Harshman, N. L and Ranade, K. S. Observables can be tailored to change the entanglement of any pure state. *Phys. Rev. A* 84 (2011), 012303.

35. Thirring, W., Bertlmann, R. A., Köhler, P. and Narnhofer, H. Entanglement or separability: The choice of how to factorize the algebra of a density matrix. *Eur. Phys. J. D* 64 (2011), 181.

36. Aspect, A., Grangier, P. Roger, G.: Experimental tests of realistic local theories via Bell's theorem. Phys. Rev. Lett. 47, 460–463 (1981).

37. Georgescu, I. How the Bell tests changed quantum physics. *Nat. Phys.* 3 (2021), 374–376.
38. Turing, A. On computable numbers, with an application to the Entscheidungsproblem. *Proc. London Math. Soc. Ser. 2* 42 (1937), 230–265.
39. Hopcroft, J. and Ullman, J. *Introduction to Automata Theory, Languages, and Computation.* Addison-Wesley, Boston MA, (1979).
40. Rice, H. G. Classes of recursively enumerable sets and their decision problems. *Trans. Amer. Math. Soc.* 74 (1953), 358–366.
41. Bohr, N. The quantum postulate and the recent development of atomic theory. *Nature* 121 (1928). 580–590.
42. Zurek WH Quantum Darwinism. Nat Phys 2009; 5:181–188.
43. Blume-Kohout, R. and Zurek, W. H. Quantum Darwinism: Entanglement, branches, and the emergent classicality of redundantly stored quantum information. *Phys. Rev. A* 73 (2006), 062310.
44. Korbicz, J. K. Roads to objectivity: Quantum Darwinism, spectrum broadcast structures, and strong quantum Darwinism – A review. *Quantum* 5 (2021), 571.
45. Zurek WH Decoherence, einselection, and the quantum origins of the classical. Rev Mod Phys 2003; 75:715–775
46. Ollivier H, Poulin D, Zurek WH Environment as a witness: selective proliferation of information and emergence of objectivity in a quantum universe. Phys Rev A 2005; 72:042113,
47. Fields, C. Quantum Darwinism requires an extra-theoretical assumption of encoding redundancy. *Int. J. Theor. Phys.* 9 (2010), 2523–2527.
48. Barrett, J, Hardy, L. and Kent, A. No-signaling and quantum key distribution. *Phys. Rev. Lett.* 95 (2005), 010503.
49. Grinbaum, A. How device-independent approaches change the meaning of physical theory. *Stud. Hist. Phil. Mod. Phys.* 58 (2017). 22–30.
50. Popescu, S.: Non-locality beyond quantum mechanics. Nature Phys. 10, 264–270 (2014).
51. Fuchs, C. A. On Parcipatory Realism. In (I. T. Durham and D. Rickles, eds.) *Information and Interaction: Eddington, Wheeler, and the Limits of Knowledge* pp. 113–134. Springer, Berln, New York, 2017.
52. Wheeler, J. A. The computer and the universe. *Int. J. Theor. Phys.* **1982,** *21,* 557–572.
53. Knill, E.; Laflamme, R. Theory of quantum error-correcting codes, Phys. Rev. A 1977; 55: 900–911.
54. Kribs, D., Laflamme R. and Poulin, D. Unified and generalized approach to quantum error correction. *Phys. Rev. Lett.* 94 (2005), 180501.
55. Laflamme, R., Miquet, C., Paz, J. P. and Zurek, W. H. Perfect quantum error correcting code. *Physical Rev. Lett.* 77(1) (1996), 198–201.
56. Lulli, M., Marcianò, A. and Piscicchia, K. Stochastic Ricci Flow dynamics of the gravitationally induced wave-function collapse, [arXiv:2307.10136 [gr-qc]]
57. Bennett, C. H., DiVincenzo, D. P., Smolin, J. A. and Wootters, W. K. Mixed state entanglement and quantum error correction. *Phys. Rev. A* 54 (1996), 3824.
58. Einstein, A. Quanten-Mechanik und Wirklichkeit. *Dialectica* 2 (1948), 320–324 (Translation by D. Howard, "Einstein on locality and separability," *Stud. Hist. Phil. Mod. Phys.* 16 (1985), 171–201.)
59. Shor, P. W. Scheme for reducing decoherence in quantum computer memory. *Phys. Rev. A* 52 (1995), R2493(R).
60. Shor, P. W. Polynomial time algorithms for prime facorization and discrete logarithms on a quantum computer. *SIAM J. Computing* 26(5) (1997), https://doi.org/10.1137/WS0097539795293172
61. Almheiri, A., Dong, X. and Harlow, D. Bulk locality and quantum error correction in AdS/CFT.*JHEP* 4 (2015), 133.

62. Maldecena, J. The large N limit of superconformal field theories and supergravity. *Adv. Theor. Math. Phys.* 2(4) (1998), 231–252.
63. Bain, J. Spacetime as a quantum error correcting code? Stud. Hist. Phil. Sci. B 2020; 71: 26–36.
64. Fields, C., Glazebrook, J. F. and Marcianò, A. Communication protocols and QECCs from the perspective of TQFT, Part II: QECCs as spacetimes. Fortschritte der Physik 2024; 72: 202400050.
65. Harlow, D. TASI lectures on the emergence of bulk physics in AdS/CFT. *Proc. Sci.* 305 (2018). 002.
66. Ryu, S. and Takayanagi, T. Holographic derivation of entanglement entropy from AdS/CFT. *Phys. Rev. Lett.* 96 (2006), 181602.
67. Lee, J.-W. Quantum entanglement from the holographic principle, Preprint arXiv:1109.3542v1, 2011.
68. Su, X.-L., Hamma, A. and Marcianò, A. Scrambling power of soft photons, to appear on arXiv (2023).
69. Su, X.-L., Hamma, A. and Marcianò, A. On the irrelevance of the scrambling power of gravity for black hole radiation: a way out from the information loss paradox, to appear on arXiv (2023).
70. Baez, J. C. Spin networks in gauge theory. *Adv. Math.* 117(2) (1996), 253–272.
71. Marcianò, A., Chen, D., Fabrocini, F., Fields, C., Greco, E., Gresnigt, N., Jinklub, K., Lulli, M., Trezidis, K. and Zappla, E. Quantum neural networks and toplogical quantum field theories. *Neural Networks* 153 (2022), 164–178.
72. Beer, K., Bondarenko, D., Farrelly, T., Osborne, T. J., Salzmann, R., Scheiermann, D. and Wolf, R. Training deep quantum neural networks. *Nature Commun.* 11(1) (2020), 1–6.
73. Zhang, C., Bengio, S., Hardt, M., Recht, B. and Vinyals, O. Understanding deep learning requires rethinking generalizations. In *Proc ICLR*, Toulon, CS, Fr. 2017
74. Marcianò, A., Zappala, E., Torda, T., Lulli, M., Giagu, S., Fields, C., Chen, D. and Frabrocini, F. Deep neural networks as the semi-classical limit of topological quantum neural networks: The problem of generaliztion. arXiv:2210.13741v2 [quant-ph] (2024).
75. Pastawski, F., Yoshida, B., Harlow, D. and Preskill, J. Holographic quantum error-correcting codes: Toy models for the bulk/boundary correspondence. *JHEP* 6 (2015), 149.
76. Maldacena, J. and Susskind, L. Cool horizons for entangled black holes. Fortschr. Physik 61 (2013), 781–811.
77. Susskind, L. Copenhagen vs Everett, teleportation, and ER=EPR. *Fortschr. Physik* 64 (2016), 551-564.
78. Fields, C., Glazebrook, J. F., Marcianò, A., and Zappala, E. (2025) ER = EPR is an operational theorem. Physics Letters B 860, 139150.
79. Dai, D.-C., Minic, D., Stojkovic, D. and Fu, C. Testing the ER=EPR conjecture. *Phys. Rev. D* 102 (2020), 066004.
80. Ji, Z., Natarajan, A., Vidick, T., Wright, J., Yuen, H. MIP* = RE. Preprint arxiv:2001.04383 (2020). Summary: *Comms. ACM 64*(11) (2020), 131–138.
81. Deutsch, D. Quantum mechanics near closed timelike lines. *Phys. Rev. D* 44(10) (1991), 3197–3217.
82. Culf, E. and Mastel, K. RE-completeness of entangled constraint satisfaction problems. arXiv:2410.21223v1 [quant-ph], 2024.
83. Qin, M. and Yao, P. Decidability of fully quantum nonlocal games with noisy maximally entangled states. In (K. Etessami, U. Fiege and G. Puppis, eds.) *50th Int. Coll. on Automata, Languages and Programming (ICALP) 2023)*, Art. 97, pp. 97:1–97 20 Leibniz Int. Proc. in Informatics Schloss Dagstuhl - Leibniz-Zentrum für Informatik, Dagstuhl Publ., Germany.
84. Vidick, T. MIP*=RE: A negative resolution to Connes' embedding problem and Tsirelson's problem. *Proc. ICM July 6-14 2022* 6, 4996–5025. EMS Press, Zurich CH, 2022.

85. Cleve, R., Hoyer, P., Toner, B. and Watrous, J. Consequencies and limits of nonlocal strategies. In *Proceedings. 19th IEEE Annual Conference on Computational Complexity, 2004*, pp. 236–249, 2004.

86. Ben-Or, M., Goldwasser, S., Kilian, J. and Wigderson, A. Multi-prover interactive proofs: how to remove intractability assumptions. In *Proc. 20th Annual ACM Symposium on Theory of Computing, STOC'88*, pp. 113–131, 1988.

87. Feige, U. and Lovász, L. Two prover one round proof systems. Their power and their problems. In *Proceedings of the 24th annual ACM Symposium on Theory of Computing*, pp. 733–744, ACM, 1992.

88. Fields, C., Glazebrook, J. F., Marcianó, A. and Zappala, E. Whether a quantum computation employs nonlocal resources is operationally undecidable. EPL 151 (2025) 48001.

89. Hofer-Szabó, G. EPR correlations, Bell inequalities, and common-cause systems. *Probing the Meaning of Quantum Mechanics*. Singapore, World Scientific, 2014, pp. 263–277. https://doi.org/10.1142/9789814596299_0013

90. Bell, J. S. Against 'measurement'. Physics World 1990; 3(8): 33–41.

91. Luminet, J.-P. Closed timelike curves, singularities and causality: A survey from Gödel to chronological protection. *Universe* 7(1) (2021), 12.

92. Morris, M. S., Thorne, K. and Yurtsever, U. Wormholes, time machines, and the weak energy condition. *Phys. Rev. Lett.* 61 (1988), 1446.

93. Hawking, S. W. Chronology protection conjecture . *Phys. Rev. D* 46 (1992), 603.

94. Brun, T. Computers with closed timelike curves can solve hard problems. *Found. Phys. Lett.* 16 (2003), 245–253.

95. Bacon, D. Quantum computational complexity in the presence of closed timelike curves. *Phys. Rev. A* 70 (2004), 032309.

96. Aaronson, S. and Watrous, J. Closed timelike curves make quantum and classical computing equivalent. *Proc. R. Soc. A* 465 (2009), 631–647.

97. Mastel, K. and Slofstra, W. Two prover perfect zero knowledge for MIP*. In *Proc. 56th Annual ACM Symp. on Theory of Computing* (New York, NY, USA, 2024), STOC 2024, Association for Computing Machinery, pp. 991–1002.

98. Moore, E.F. Gedankenexperiments on sequential machines. In Autonoma Studies; Shannon, C.W., McCarthy, J., Eds.; Princeton University Press: Princeton, NJ, USA, 1956; pp. 129–155.

99. Dietrich, E., Fields, C.: Equivalence of the Frame and Halting problems. Algorithms 13, 175 (2020).

Active Inference and Self-organizing Systems

8

As we have emphasized in previous chapters, particularly in Chaps. 4 and 6, the FEP is a formal, completely general, scale-free framework for describing interactions between physical systems in cognitive terms. It states that any system S that interacts with its environment E weakly enough to maintain its identifiability over time possesses an MB that separates its, S's, internal states from the internal states of E, and that any such S asymptotically minimizes the VFE measured at its MB by using its GM to generate probabilistic beliefs about the consequences of actions. We also recall how the FEP, as the cornerstone of active inference, states that any system with a NESS solution of its density dynamics will act so as to maintain its state in the vicinity of its NESS, or equivalently, requires that almost all paths through the joint space that begin in $S(E)$, remain in $S(E)$ [1]. Hence any system compliant with the FEP can be described as engaging, at all times, in active inference: a cyclic process in which the system observes its environment, updates its Bayesian beliefs (i.e., posterior or conditional probability densities) over future behaviors, and acts upon its environment both to test its predictions and to bring its environment's state into closer compliance with them. This process can be manifest in several ways, for instance:

- perception, insight, and (positive) engagement in some action [2, 3];
- implementing a (multi-level) policy, agenda, or goal attainment, and the fulfilment of any of these [2, 4–6];
- learning through interaction with an environment, and/or the recognition of a fact (as learned) [7, 8];
- reaching a decision (intended as optimal one) [4, 6, 9];
- the eventual 'pay-off' for a game, in a general sense (e.g. viewing any physical interactions as a game) [10].

© The Author(s), under exclusive license to Springer Nature Switzerland AG 2026   135
C. Fields and J. Glazebrook, *Distributed Information and Computation in Generic Quantum Systems*, Synthesis Lectures on Engineering, Science, and Technology,
https://doi.org/10.1007/978-3-031-97263-8_8

- in an asymptotic limit, approaching entanglement with E in accord with the principle of unitarity/conservation of information [11].

The first four of these are evident in any "cognitive" systems, including living organisms and autonomous robots engineered for reconstructing and predicting environmental experiences [3, 7, 9, 12].

Understanding how active inference drives successful behavior, and how it is implemented by some behaving system, are thus of paramount importance, particularly in the presence of situation-meaningful complexity, and robustness when confronted with varying and challenging environmental conditions. We will see in what follows how planning and control can be represented in terms of hierarchical inference [13–15], and how the examples listed above can be understood in the context of learning, motivated control, etc.

We can turn all of this around from the 'agent', and consider matters from the 'environmental' point of view. We can interpret 'environment', as a general holistic medium, in a number of ways; for instance: social or geographical (environment), atmospheric/climatic, biochemical, action performing entities, in particular, those environments incorporated and regulated by models of a Markov decision process. The corresponding interactions can be generally, and conveniently, explained in terms of the MB, which as we pointed out, defines and delineates the boundaries of the system. In this way, the MB sees the environment as external states reciprocating with blanket sensory states which feed into the agent's (internal) states, and in turn, these latter states reciprocate with the blanket active states which feed back into the environment, hence completing the stages of an agent's response to its environment (see e.g. [16], and specifically for neuronal dynamics [17]).

8.1   The Environment as an Active Agent

Previous chapters have emphasized the universality of the FEP as a description of physical systems that persist, i.e. remain distinct from their environments, over macroscopic time. Decomposing a Hilbert space as $\mathcal{H}_U = \mathcal{H}_S \otimes \mathcal{H}_E$ defines two physical systems, S and E. If one of these systems persists, clearly the other does too. The "environment" E of any system S that is taken to be "of interest" is, therefore, every much a subject of the FEP as S is, and can, in particular, be regarded as an active inference agent. The Conway-Kochen theorems [18, 19] render any such environment a "free" agent, e.g. free to choose its own QRFs as discussed in Chap. 2.

The sense of the "environment" as an agent becomes compelling when a generic physical interaction can be interpreted as a "game" in the broadest sense, as comprehensively treated in [10]. The systems S and E defined by a Hilbert-space decomposition are regarded as the "players" of the interaction-as-game. Milnor, for example, introduced the idea of a "game against Nature" where the environment E plays the role of "Nature" [20]; E can, for example, be interpreted as a collection of resources, e.g. thermodynamic free energy, stigmergic memory, etc. with some degree of stochasticity, as discussed earlier in Chap. 3. Then as before we can write the interaction between S and E as a Hamiltonian $H_{SE} = H_U - (H_S + H_E)$, where H_U, H_S, and H_E are the internal or self-interactions of U, S, and E, respectively. The boundary $\mathcal{B}$ is the "board" (or "media", or "channel") on, or through which the game is played. This channel can be any information-encoding space; for instance, the internet in the case of video games, or a private channel in a quantum cryptography setting where an eavesdropper effectively plays a game with a subject who has assumed total privacy [21]. Given $\mathcal{B}$, we can describe the moves and the strategies that drive them. Each move has two components: first S (E) prepares each qubit on $\mathcal{B}$ in some state, after which E (S) measures each qubit. As a physical system that provides a fixed, re-usable standard for measurements in the face of an environment, the QRF meets the task, as we will further specify in Sect. 8.1.2 below.

8.1.1 Symmetry and Symmetry Breaking on the Boundary

Let us now revisit the setting of previous chapters, where we consider two agents A and B separated by a holographic screen or MB, and whose dynamics are represented by QRFs. The fundamental prescription, within a purely topological setting, is given by (see also [5, II] and references therein):

- We can regard A and B as separated, and determined by independent measures. They are separated by—and interact via—a holographic screen $\mathcal{B}$ that can be represented, without loss of generality, by an array of N non-interacting qubits, where N is the dimension of H_{AB}.
- A and B can be regarded as exchanging finite N-bit strings, each of which encodes one eigenvalue of H_{AB}.
- A and B have free choice of basis for H_{AB}, corresponding to a free choice of local frames at $\mathcal{B}$; for instance, a free choice for each qubit q_i on $\mathcal{B}$, of the local z axis, and hence the z-spin operator s_z that acts on q_i.
- A choice of basis corresponds to choosing the zero-point of total energy by each of A and B. The systems A and B are, therefore, in general at informational, but not at thermal equilibrium.
- As A and B must obtain from B or A, respectively, whatever thermodynamic free energy is required, by Landauer's principle (see Chap. 3) to fund the encoding of classical bits on $\mathcal{B}$ (as well as any other irreversible classical computation), A and B must each devote some sector F of $\mathcal{B}$ to free-energy acquisition. The bits in F are "burned as fuel", and thus do not contribute input data to computations. Waste-heat dissipation by one system is free energy acquisition by the other. The free-energy sectors F_A and F_B of A and B need not align as subsets of qubits on $\mathcal{B}$; that is, qubits that A regards as free-energy sources may be regarded by B as informative outputs and vice-versa.

- The actions of the internal dynamics H_A and H_B on $\mathscr{B}$ can be represented by A- and B-specific sets of QRFs, each of which both "measures" and "prepares" qubits on $\mathscr{B}$. Each QRF acts on the qubits in some specific sector of $\mathscr{B}$, breaking the permutation symmetry in the expression for the interaction Hamiltonian H_{AB}. Only QRFs acting on sectors other than F implement informative computations; we will therefore restrict attention to these QRFs.
- Each "computational" QRF can, without loss of generality, be represented by a CCCD, as in Chap. 3, recalling that the apex of each such CCCD is, by definition, both the category-theoretic limit and colimit of the I/O classifiers that formally correspond to the operators M_i^k as in the expression for H_{AB} (again, we recall the blueprint of a variational autoencoder as an exemplar).

Note that the holographic screen $\mathscr{B}$ functioning as an MB separating A from B, can be seen as having an 2^N-dimensional, N-qubit Hilbert space $\mathcal{H}_{q_i} = \prod_i q_i$. While $\mathcal{H}_{q_i}$ is strictly ancillary to $\mathcal{H}_U = \mathcal{H}_A \otimes \mathcal{H}_B$, the classical situation can be recovered in the limit in which the entanglement entropies $\mathcal{S}(|A\rangle), \mathcal{S}(|B\rangle) \to 0$ by considering the products $\mathcal{H}_A \otimes \mathcal{H}_{q_i}$, and $\mathcal{H}_B \otimes \mathcal{H}_{q_s}$, to be "particle" state spaces for A and B, respectively. In this classical limit, the states of $\mathcal{H}_{q_i}$ become the blanket states of an MB. This characterizes the function of the MB as a classical information channel. This is specified by the classical limit of $\mathscr{B}$ [22–24], and crucially, *if agents/systems A and B are separated by a finite $\mathscr{B}$, then entanglement entropy of the joint state $\mathcal{S}(|AB\rangle) = 0$* [25, Lemma 1].

Now let us consider some consequences of a QRF deployed by one of the agents, A say; that is, the QRF is implemented by H_A as corresponding to a set of observables held fixed, while other observables are allowed to vary freely [24, 26]. The associative groupings of the operators M_i^A in H_{AB} can be seen to be independent of the associative groupings of the M_i^B. Hence, the choices of a QRF by A have no bearing on choices of a QRF by B, or vice-versa (equivalently, swapping the labels A and B has no effect on the expression for H_{AB}). Technically, this "free choice" of QRFs corresponds to the absence of a *superdeterminist correlations* between A and B. Such correlations implement entanglement [27, 28], and so are forbidden if $|AB\rangle = |A\rangle|B\rangle$ (see [25] for further discussions). Notably, communication protocols that employ shared entanglement depend on shared QRFs (see Chap. 7 and [29]).

In recalling the pointer and reference states employed for QRFs in Chap. 2 (see also [25, 26]), the assumption of classical communication is, effectively, the assumption of a preferred pointer measurement that returns the content of the communicated message without requiring prior identification, via a separate measurement, of the physical medium; that is, of the QRF, via which the message has been transmitted. An explanation is implicit in the Alice and Bob scenario: Alice, in other words, does not have to identify Bob to receive his message, just as Wigner does not, in his famous thought experiment, have to identify his friend to receive his friend's observational outcomes [30]. Hence to assume classical communication is to assume an a priori shared QRF [24]. Crucially, *this breaks the free-choice symmetry across $\mathscr{B}$*, when the latter is considered a qubit array as seen in [25, Fig. 3]. The assumption of classical communication is the assumption that A and B use identical z-axis QRFs on a subset of qubits. For further the details and further consequences of these and other results, see [25, Sects. 3.2–3.5].

In summary, [25], demonstrates a generalization of the holographic principle in which interactions H_{AB} between finite quantum systems A and B that maintain a separable joint state, are represented as exchanges of information across the holographic screen (MB) $\mathscr{B}$. While the role of $\mathscr{B}$ is ancillary to the action of H_{AB}, the permutation group symmetry (that is, group-theoretically the S_N-symmetry) of $\mathscr{B}$ is broken when the internal Hamiltonian H_A is considered to engage QRFs that both identify the systems in question, and measure their states. So, now back to the environment. This symmetry breaking induces decoherence of identified systems by forcing the remaining environment to serve as both a free energy source and waste-heat sink. Observable entanglement, contextuality, and classical memory are, in this representation, logical and temporal relations between QRFs implemented by H_A. This is part of the background to the control of active inference systems pursued in [5, 31]), elements of which we will bring to the forefront in the subsections below.

8.1.2 QRF Requirements for Meaningful Interaction

We have seen how symmetries in relationship to $\mathscr{B}$ can potentially influence an agent's attainment, or non-attainment, or in some cases in some cases, a fleeting attainment to active inference. Now we will probe a little deeper and look at the symmetries of the QRFs themselves, and the ensuing dynamics thereof. The purpose of these symmetries follows from, and generalizes to an extent, classical situations as they arise, for instance, in the theory of transformations of Galilean and Special Relativity in the form of group actions.[1] However, it is only in recent years that the quantum aspect is gaining its gradual development, with the introduction of such important concepts as that of a *positive operator-valued measure* (POVM) which will receive some attention below; for surveys and background to the subject in general, see e.g. [34, 35].

Let us commence by taking some background QRF, and a quantum system X in a state s_X with respect to it. Changes of orientation of X relative to this frame are describable by a unitary representation $U_X(g)$ of an element g of a transformation group G that determines each change of orientation. Next, let Y be another system exhibited in a QRF state s_Y, also defined with respect to this QRF. The reference frame state breaks a G-symmetry, and is represented by a unitary U_Y-action acting on the Hilbert space $\mathcal{H}$ of the QRF system $\mathcal{H}_Y$. At this stage, it is necessary to de-implicate the background QRF. As shown in [35], this is achieved by de-implicating the compound system XY from the background frame. This

---

[1] Typically, these will be Lie group actions, the invariance thereof, etc. (for surveys of Lie groups and their representations, see e.g. [32, 33]).

commences by taking some state s and averaging it over all rotations $g \in G$, a locally compact group, on using $U_{XY}(g) = U_X(g) \otimes U_Y(g)^2$ then in terms of bounded linear operators, this induces a map $\mathcal{G} : \mathcal{L}(\mathcal{H}) \longrightarrow \mathcal{L}(\mathcal{H})$, the 'G-twirl' of the state s, defined by

$$\mathcal{G}(\mathsf{s}) = \int d\mu(g)\, \mathcal{U}_{XY}(g)[\mathsf{s}] = \int d\mu(g)\, U_{XY}(g)\, \mathsf{s} U_{XY}(g)^{\dagger} \tag{8.1}$$

where $\mathcal{U}_{XY}(g)[\mathsf{s}] = U_{XY}(g)\, \mathsf{s} U_{XY}(g)^{\dagger}$ is the unitary map of the left-action of G and $d\mu(g)$ denotes the latter's (normalized) invariant Haar measure.[3]

An *encoding map*:

$$\mathcal{E}_{\mathsf{s}_Y}(\mathsf{s}_X) = \mathcal{G}_{XY}(\mathsf{s}_X \otimes \mathsf{s}_Y) \tag{8.2}$$

is a means of seeing how a quantum system X, defined relative to a background frame can be de-implicated from this background by introducing a QRF Y. This produces a G-invariant state $\sigma_{XY} = (E)_{\mathsf{s}_Y}(\mathsf{s}_X)$; a *relational encoding* of s_X using s_Y, implementable by applying (8.1) to $\mathsf{s}_X \otimes \mathsf{s}_Y$ where the representation on $\mathcal{H}_X \otimes \mathcal{H}_Y$, is given by $U_X(g) \otimes U_Y(g)$. The process of unravelling all this, that is, to recover s_X with respect to the background frame is attainable via a *recovery map* [35] (and references therein):

$$\mathcal{R}(\sigma_{XY}) = $$
$$D_{XY} \int d\mu(g)[U_X(g^{-1}) \otimes \langle g|_Y]\sigma_{XY}[U_X(g^{-1})^{\dagger} \otimes |g\rangle_Y] \tag{8.3}$$

that yields a state s'_X on $\mathcal{H}_X$. Significantly, (8.3) describes the measurement of a QRF on system X against a background reference frame formed by a POVM with elements proportional to projections onto the states $|g\rangle$ on the QRF. The quantum statistics of a relational measurement of two QRFs Q_A and Q_B are determined by a relational POVM $\{E_h : h \in G\}$ permitting calculation of probabilities of h outcomes given the input states. Accordingly, a measurement can be defined as a POVM [35] when it is possible to construct a family of trace-decreasing, completely positive maps $\mathcal{F}^h_{AB}$ associated with POVM elements to determine the post-measurement state for a given outcome h. Towards establishing this, many technical details employing the calculus of projection operators are exhibited in [35, III], following which is an account of performance and decoherence under change of QRFs, and how the results with the consequences of these can be interpreted.

Much of this suggests that for defining a QRF, we include admittance of a POVM covariant under the representation U of G.[4] As pointed out in [34, 3.1], there are physical motivations for this requirement. An example is the presence of a non-trivial observable paired with a state with the purpose of inducing frame 'orientations' probabilistically, with the

---

[2] That is, the unitary representation of G on the combined tensor product $\mathcal{H}_X \otimes \mathcal{H}_Y$.

[3] For G-invariant states [35] speaks of 'G-twirling invariant states', which are independent of choice of a background QRF.

[4] *Covariance* here applies to a function or operator when the latter is invariant under a linear transformation of variables.

covariance property reflecting the properties for classical co-ordinate changes, that include active-passive duality. Accordingly, a QRF $\mathcal{R}$ is formally definable as a system of covariance $\mathcal{R} = (U_\mathcal{R}, \mathsf{E}_\mathcal{R}, \mathcal{H}_\mathcal{R})$ on a homogeneous (left) G-space $\Sigma_\mathcal{R}$.[5] Here, when G is understood, $\mathsf{E}_\mathcal{R}$ is the 'frame observable' once $U_\mathcal{R}$ and $\mathcal{H}_\mathcal{R}$ are also understood [34, Definition 3.1]. This formal definition of a QRF specifies its *operational nature*, properties of which are already implicit in the general formulation of Chap. 2, and along with that of a POVM, has been recently exhibited in [36]. We refer to [34] for special types of these operational QRFs, and the related theory of operator-algebras which determine and control the dynamics of the resulting systems.

Box 8.1: Group actions as QRF morphisms

Let $|\psi\rangle^A_B$ denote the state of a system **B** relative to that of **A**. If $|\psi\rangle^A_B = |0\rangle^A_A \otimes |\psi\rangle^A_B$, then establishing the state $|\psi\rangle^B_A$ (i.e. that of **A** relative to **B**) is a matter of transforming QRFs by the action of a symmetry group. This question is treated in depth in [37], in accordance with the principles of relational physics, for both reversible and irreversible changes of reference, and conditions for implementing unitary transformations of these. Such a study opens up the door for a prospective (symmetry) category of QRFs, where Objects are the QRFs themselves, and Arrows the (associative, with identity) transformations between QRFs. Such a category could be studied in relationship to that of CCCDs [23], and the various concepts discussed in Appendix Sect. D.2.

8.1.3 How the Environment Enables Learning

Let us now return to the FEP. As a least-action principle, it is fundamentally a statement about those paths followed by the joint system U passing through its state space. We recall that the classical FEP is amenable to a path-integral formulation [1] that expresses the expected value of any observable (functional) $\Omega[x(t)]$ of paths $x(t)$ through the relevant state space as [38, Eq. 6]:

$$\langle \Omega[x(t)] \rangle = \int dx_0 \int d[x(t)] \Omega[x(t)] p(x(t)|x_0) p_0(x_0) \tag{8.4}$$

where x_0 is the initial state, and $p(x(t)|x_0)$ is the conditional probability of the path $x(t)$. This is generalized in quantum theory by effectively replacing $\Omega[x(t)]$ with an automorphism of the relevant Hilbert space and $p(x(t)|x_0)$, with an amplitude for $x(t)$ given the initial state x_0. This connects with the formulation of the TQFT discussed in Chap. 7 (see also [23]).

As we pointed out earlier, when S encodes a bit string on $\mathcal{B}$ by preparing each of the q_i in some particular state, it must select a local $+z_i^S$ QRF for each of the M_i^S, and hence for each of the q_i. When the environment E subsequently reads a bit string from $\mathcal{B}$, it must

---

[5] That is, a space of the form K/H, where typically in the category of Lie groups, K, H are certain closed, connected Lie groups on which G acts on the left (see e.g. [32, 33]).

also select a local $+z_i^E$ QRF for each of the q_i. If S and E are to remain separable, these choices of local QRFs, which correspond to choices of basis $|i\rangle$ in the expression for the interaction Hamiltonian H_{AB}, must be made independently or "freely" [23]: if S's choice of basis depends on E's or vice-versa, they are entangled. Viewing $\mathscr{B}$ as a communication channel, an independent choice of basis by both S and E can be viewed as introducing noise into the communication; in the extreme case of S choosing $+z_i^S = -(+z_i^E)$ for q_i, S will observe E's encoded bit as being flipped. As U is isolated, there is no classical source of noise in the system; the "noise" due to differences in QRF/basis choice between S and E, is purely quantum. Note that noise caused by distinct QRF choices generates statistical surprise. Take for instance, when playing some game, surprise can be induced when one's opponent suddenly switches to a different strategy.

This particular example is of the type that can be fed into the TQFT picture (as in Chap. 7) towards control flow of active inference systems; mainly in terms of classical and quantum channels as we had in discussed Chap. 7. So it is fitting to recall the descriptive mechanism for implementing this, i.e. Diagram (7.5), which we replicate here for convenience:

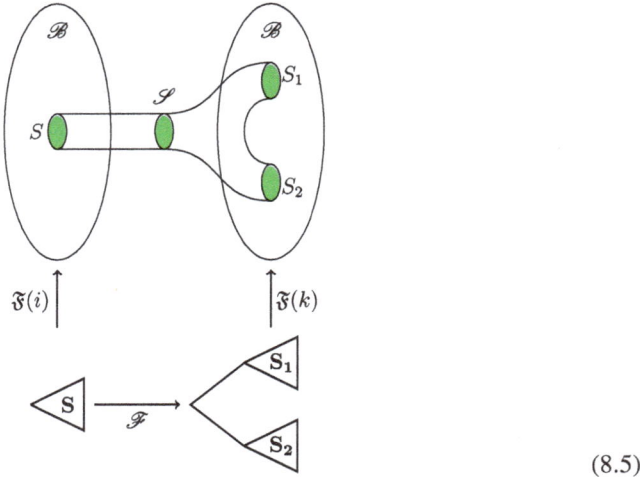

$$(8.5)$$

We recall from Chap. 7 and [23, Theorem 1.1]: *for any morphism $\mathscr{F}$ of CCCDs in* **CCCD**, *there is a cobordism $\mathscr{S}$ such that a diagram of the form of Diagram (8.5) commutes.* The main thrust of this result is that it applies distributed information flow in the formulism of a global workspace to any sequential measurement. It is particularly relevant to any active inference framework respecting an environment in informational terms as it applies to measurements of a sector X, followed by measurements of the associated memory sector Y, or vice versa (*cf.* Diagram (7.8)):

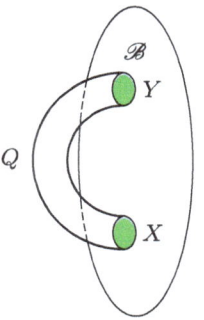

$$(8.6)$$

As demonstrated in [5], by assuming for convenience that $\dim(X) = \dim(Y)$, we can consider a composite operation $Q = (\overrightarrow{Q}, \overleftarrow{Q})$, where $\overrightarrow{Q} = Q_X Q_Y$ and $\overleftarrow{Q} = Q_Y Q_X$, is then a pair of QRF sequences that can be identified with TQFTs that measure and record an outcome, mapping $\mathcal{H}_X \rightarrow \mathcal{H}_Y$, and dually use an outcome read from memory to prepare a state, mapping $\mathcal{H}_Y \rightarrow \mathcal{H}_X$, respectively as in Diagram 8.6 above.

Now recall Diagram (2.4), which we replicate as Diagram (8.7) below. Note that, technically, our identification of QRFs as essentially internal TQFTs, is conducive to a more general analysis of information exchange between multiple QRFs deployed by a single system (such as A in (2.4)) towards (information) control flow. Further, since all QRFs act on the boundary $\mathcal{B}$, information exchange between QRFs requires a channel that traverses $\mathcal{B}$. Any such channel is itself a QRF, one deployed by B, as shown in Diagram 8.7 below. On considering A to comprise two observers, one deploying Q_1 and the other deploying Q_2, that interact via a LOCC protocol provides an example [39] (see also see Chap. 7 and [29]):

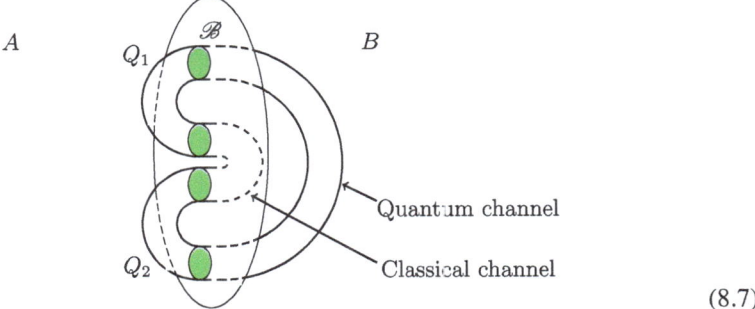

$$(8.7)$$

In an LOCC protocol, one channel is considered 'classical' while the other is considered 'quantum'. However, this language masks the fact that both channels are physical. As pointed out in [27] (and see Chap. 7), all media supporting classical communication are physical, and interactions with these media are always local measurements, or preparations. Therefore, the two channels in an LOCC protocol are physically equivalent. Both are seen as TQFTs implemented by B, although each posses distinct conventional semantics.

Diagram (8.7) above can also represent externally-mediated communication between any two functional components of a system, exemplified by macromolecular pathways within a cell, or functional networks within the brain (cf. [17]). In [5], we showed that whenever Q_1 and Q_2 are deployed by distinct—technically, separable or mutually decoherent—"observers" or "systems," they fail to commute, i.e., the commutator $[Q_1, Q_2] = Q_1 Q_2 - Q_2 Q_1 \geq h/2$, where again h is Planck's constant. As shown in [40, Sect. 7] using the CCCD representation, non-commutativity of QRFs induces quantum contextuality (see also [41], Chap. 5, and references therein) where dependence of measurement results on non-local hidden variables that characterize the measurement context.

In this present framework of active inference, such hidden variables characterize the action of H_B on $\mathscr{B}$, affecting what system A will observe next in every cycle of the A-B interaction. In principle, such context dependence can be captured classically if sufficient measurements of the context can be implemented (see [40] and references within). But these measurements would necessitate access to all of B. The presence of active inference stresses the point: the existence of an MB prevents such access; in the present picture this means that A has access to B only via $\mathscr{B}$. The finite energetic cost of measurement, and consequent requirement for a thermodynamic sector F, prevents measurement, even of all of $\mathscr{B}$, by any finite physical system. This is why we expect physical systems, including all biological systems, to employ only local context-dependent control to switch between mutually non-commuting (sets of) QRFs. A discussion of how context switches as implemented by a QRF induce evolution, development and learning, is presented in [42].

Box 8.2: Minimal Physicalism

Minimal Physicalism (MinPhy) was introduced in [42] as a methodology for understanding the experience of generic physical systems that depends only on the minimal assumptions that have been made in this and earlier chapters. It provides a conceptual bridge between the information-flow formalism developed here and the growing research areas of basal cognition and multi-species consciousness. The fundamental idea of MinPhy is that any system, at any scale, *experiences* the data written on its boundary/MB, as processed by the QRFs that it deploys. This idea has been implicit in much of the discussion in this book. Briefly,

1. MinPhy is completely scale-free, applying in the same form to interactions between molecules, cells, tissues, organisms, social groups, or ecosystems and their respective environments. Hence it predicts common mechanisms that can be probed empirically at any scale. It makes no assumptions about the structure or dynamics of the environment, at any scale, beyond it being a physical system.

2. MinPhy requires every property of either itself or its environment to which an organism, or other living system, is differentially sensitive to be specified explicitly in terms of the information processing—the QRFs—employed by the organism or system to detect and respond to that property. This includes properties often taken

for granted, such as the existence of external objects or their embedding in three dimensional space, and applies regardless of whether detection and/or response is recorded to memory or "reportable" via any specific behavior or assay.

3. MinPhy treats information and energy as formally equivalent, explicitly enforces conservation of energy, and hence requires the thermodynamics of classical computation to be represented explicitly. It requires the thermodynamic cost of memory to be accounted for at every scale, and automatically enforces a metabolism-cognition trade-off.

4. MinPhy associates meaning with *actionability* in the spirit of Bateson's definition of meaning as a "difference that makes a difference" [43] (see also [44, 45]).

As an example, let us consider the *modus operandi* of a cell membrane (in eukaryotic cells, also any organelle membrane) which implements the MB, the blanket states of which comprise the states of embedded receptors as well as other physiological, e.g. mechanical or electrical, properties of the membrane. The existence and function of the MB provides the cell with a well-defined state, conditionally independent of its environment. Information transfer across the MB enables vital homeostatic and allostatic processes to the extent that the transferred information is meaningful for—actionable by—the cell. Any collection of microsystems that individually minimize VFE at their boundaries for which a collective boundary can be defined constitutes, in MinPhy, a macrosystem that experiences the information transferred across the collective boundary; a network of neurons or other mutually-communicating cells, for example, constitutes a MinPhy agent. Such systems engage in active inference and hence in biological self-organization; this collective activity of "parts" at any scale drives the evolution of larger-scale systems with higher complexity [46]. MinPhy provides a natural approach to the origin of life [47], and furthermore suggests that "life" is a universal property of self-organizing systems [48].

8.1.4   Tensor Networks as Control Systems

We demonstrated in [31] that a control flow of active inference in any systems can always be formally described as a *tensor network* (TN), as summarized earlier in Box 6.1. Informally, a TN arises from a dataset (quantum, or otherwise) as a factorization of a tensor (i.e., high-dimensional matrix) operator into multiple component tensor operators that are pairwise contracted on shared degrees of freedom. This factorization is seen as approximated by a composition of smaller multilinear maps.

Tensor Networks provide representations of generic networks prescribed by nodes and links that are susceptible to training, so are notably instrumental in classical-to-quantum Machine Learning (ML), as surveyed in [49], and for Deep Learning (DL) [50]. From the perspective of quantum information, [51] points out that a TN is a many-body wavefunction obtained by gluing together few-body quantum states which comprise the tensors, and notes that creating a TN is rather like connecting a number of computers to the Internet. Communicating via email does not entail computers sending messages to each other, but signals are instead sent to a number of servers which mediate the communication. Regards causality, in the context of DL, any trained network implements a causal model. For the sake of algorithmic effectiveness, one precondition is to adopt the *causal faithfulness condition*, that along with the *causal Markov condition* implies that any causal Bayesian network satisfying these conditions admits a unique MB [52–54].[6] Below we will discuss their relevance to the TQFTs of Chap. 7.

In [5, 31], towards a framework for control of active inference systems through distributed information, we brought to the table the question of whether, and to what extent, a cognitive system entailing active inference, can impose a decompositional or mereological (i.e., part-whole) structure on its environment. The structures in question naturally invoke a notion of locality, and hence of geometry. In fact, the geometry of spacetime itself has been described as a particular TN, in terms of a *multiscale entanglement renormalization ansatz* (MERA) (see e.g. [56–58]), in turn suggesting a deep link between the control flow in systems capable of observing spacetime (i.e., capable of implementing internal representations of spacetime), and the deep structure of spacetime as a physical construct itself. A general result in this direction is [31, Theorem 1]:

Theorem 8.1.1 *A system A exhibits non-trivial control flow if, and only if, its control flow can be represented by a TN.*

In this respect, control flow can said to be trivial when a systems deploys only a single QRF. This means that systems that deploy only a single QRF effectively "do the same thing", regardless of context. Hence, they do not qualify as "systems of interest" in the sense used here. Consequences of Theorem 8.1.1 are:

(1) *Decoherent reference sectors exist on a boundary $\mathscr{B}$, if and only if control flow can be implemented by a TN* [31, Corollary 1]. This follows from the fact that decoherence between sectors requires independently-deployable, non-commuting QRFs. This requires a control structure that factors, and hence by the above main result it requires a TN. Conversely, a TN factors the control structure, making QRFs independently deployable, which renders their sectors decoherent. Equivalently, the GM implemented by a system [12] factors if, and only if, control flow can be implemented by a TN.
(2) *The TN of any system compliant with the FEP is a decomposition of the Identity* [31, Corollary 2].

---

[6] Methods of DL, as applicable to computational biology, are reviewed in [55].

In effect, no finite physical system can measure the entire state of its boundary $\mathscr{B}$ with a single QRF, and thus no such system can simultaneously measure, and act on its entire context [31]. Any system A that deploys multiple QRFs in sequence cannot avoid contextuality due to unobservable effects, mediated by the action of the Hamiltonian H_B on the action of Q_i on the state measured by Q_j. Putting it another way, every action engaged by a putatively "interesting" system, at least in a transient sense, increases the VFE at its boundary.

Suppose we take a system A that deploys multiple, distinct QRFs, $Q_1, Q_2, \ldots Q_n$, where $n \ll N = dim(H_{AB})$. Classical control flow in A can then be represented by a matrix $\mathbf{CF} = [P_{ij}]$, where P_{ij} is the probability of the control transition $Q_i \rightarrow Q_j$. Along with such transitions there are inevitably energetic costs, which must be paid for by free energy sources within the thermodynamic sector F of $\mathscr{B}$. How Theorem 8.1.1 contributes to control of active inference schemes, will be discussed below.

8.2   Cooperation, Competition, and Negotiation

8.2.1   How the Environment Shapes Multi-component Systems

Thinking of the examples we have run through above, it is clear to say that any system with multiple perception–action (or stimulus–response) capabilities necessitates a control system that will facilitate context-guided perception and action, as well as precluding the continuous, simultaneous deployment of all available perception–action capabilities. For instance, a 'policy' from the entirety of these hosted within the system's GM, is vital for self-organization. Within the active inference framework, ranging from the microbiological to robotics, the system's internal states—hence its GM—can be read as encoding Bayesian beliefs over the causes of its sensory states; in other words, the result of the system's own actions and behavior within a given environment. Hence, planning and control, driven by policy adoption, results in inference [13–15], with the ensuing selection of an action given by the most likely 'policy' (simply put—a course of action). Often it is the case that active inference is combined with techniques of, for instance, behavior tree diagrams for the purpose of making decisions given various constraints of running-time, etc. [59]. Take for instance the dynamical strategies of some sport, where typically we have in Bayesian terms generating a GM (as studied for tennis in [60]):

$$p(\text{goal} \mid \text{movement}) = p(\text{goal}) * p(\text{movement} \mid \text{goal})/p(\text{movement}) \tag{8.8}$$

It is clearly the case that in a sport such as tennis, the 'goal' ("game, set, and match") and 'movement' (service, positioning, and dynamics of play, etc.) are environment-dependent; at the very least, the context and demarcations of the court have to be strictly observed (the weather, and presence of spectators can also be factors). Much the same applies, needless to say, to other sports.

To take a more specialized notion of 'environment', an example drawn from microbiology provides an explicit test-case. In bacteria such as *E. coli*, for instance, mutual inhibition between gene regulatory networks (GRNs) for different metabolic operons, permit the expression of specific carbon-source (for instance, sugar) metabolism pathways, importantly, only when the target carbon source is detected in the environment [61]. A feature of GRNs is that they are essentially classical in nature, even though, for instance, they can support electron-transfer which is typically seen as quantum [62, 63]. Forney-type Factor Graphs, as implemented in [64], for instance, provide another important technique when combined with messaging passing schemes in which the messaging is reflective of extant statistics towards computing elements of marginal densities, etc. The punch-line is that, under the FEP, the necessary control can be formulated inferentially; subsequently, control can be formulated in terms of the actual message passing within the factor graph (see also [65, 66]).

When the environment is modelled as a *Markov decision process* (MDP), there is a factorization of the corresponding GM that comes about as follows. At a given time t, let s_t be the state, o_t the observation, a_t the action, and s_τ the expected time at a future timestep τ. We also have the respective sequences of states, observations, and actions: $\tilde{s}$, $\tilde{o}$, and $\tilde{a}$. Further, let $P(o_t|s_t)$ be a likelihood model, given a policy π, let $P(\pi)$ be the belief over that policy, $P(s_t|s_{t-1}, a_{t-1})$ be the state transition model, and $P(a_t|\pi)$ the action given then policy π. Then the agent's GM, denoted $P(\tilde{o}, \tilde{a}, \tilde{s})$, factorizes as [9]:

$$P(\tilde{o}, \tilde{a}, \tilde{s}) = P(\pi)P(s_0)\prod_{i=1}^{T} P(o_t|s_t)P(s_t|s_{t-1}, a_t)P(a_t|\pi) \tag{8.9}$$

Similarly to before, the free energy is expressible as:

$$\begin{aligned}
\mathbf{F} &= \mathbb{E}[\log Q(\tilde{s}) - \log P(\tilde{s}, \tilde{o})] \\
&= D_{\mathrm{KL}}(Q(s\tilde{f}s)\|P(\tilde{s}|\tilde{o})) - \log P(\tilde{o}) \\
&= D_{\mathrm{KL}}(Q(s\tilde{f}s)\|P(\tilde{s})) - \mathbb{E}[\log P(\tilde{o}|\tilde{s})]
\end{aligned} \tag{8.10}$$

where $Q(\tilde{s})$ denotes an approximate posterior distribution. The explanation of terms follows as previously when interpreting the FEP, noting that the third equality, as the negated evidence lower bound, is a property of *variational autoencoders* [67] of which the CCCDs comprise suitable blueprints as we discussed in Chap. 3. Again, one of the tenets of active inference is that agents habitually visit those states for which free energy is expected to be low, using a history of prior beliefs about how much free energy is to be expended by the policy in question. This leads to the probability of choosing a probability as given by [6]:

$$P(\pi) = \sigma(-\gamma G(\pi)) \text{ where } G(\pi) = \sum_{\tau}^{T} G(\pi, \tau) \tag{8.11}$$

σ is a softmax function with precision parameter γ. The expression $G(\pi, \tau)$ is the expansion of the expected free energy of the policy $G(\pi)$, in terms of the future time-step τ under the policy π (see [9, (4)]). From this [9] deduces that action selection in active inference involves (at least in instances with a finite number of discrete pre-defined polices; cf. [7]):

(1) Evaluate $G(\pi)$ for each policy π.
(2) Calculate the belief over polices $P(\pi)$.
(3) Infer the next action using $P(\pi)P(a_{t+1}|\pi)$.

8.3 Multi-scale Competency Architecture

Michael Levin introduced the idea of a multi-scale competency architecture (MCA) to describe complex systems in which components at each level of organization "know what to do" without being micro-managed by higher-level components [47, 68, 69]. Multicellular organisms are the obvious examples: every cell in a multicellular body is competent to perform some functions—e.g. packaging oxygen into hemoglobin or filtering toxins from blood—without having to be instructed by higher-level systems. When and how fast such functions are performed are responsive to high-level modulation, but they can be performed autonomously by cells that have the competency. The use of highly-specialized, intelligent components in MCAs contrasts with the use of relatively uniform and "dumb" components in artificial computers. Here we examine control flow in such systems, from the abstract perspective of hierarchical active inference.

8.3.1 General Representation of Hierarchical Active Inference Systems

Let us recall the 'Helmholtz machine' of Dayan et al. [70] hypothesizing the brain as a statistical mechanism of inference. The FEP perspective extends this to regard the process of learning as evolution of the GM with respect to the environment in question. As before, minimization of the VFE reduces the prediction error between sensations as they are registered, and the ensuing predictions. This lends an extensive study within the framework of *associative learning*; that is, typically, a mode of learning that distinguishes between separate stimuli, learning by association, and condition-response. In the active inference picture, many models of associative learning are realized in terms of functional sensorimotor contingencies within a hierarchy of GMs of increasing depth in which degrees of attention modulate corresponding degrees of precision [71, 72].[7]

---

[7] Neurophysically, these processes are realized in such cases as the active inference of peripheral reflexes under autonomic regulation and homeostatic regulation which necessitates a continual updating of empirical priors for actions, the dynamics of which are generated by an uninterrupted flow of message passing [14, 72, 73]. In a related way, [74] show that how (canonical) NNs in which rate

In terms of models of state spaces and goal-directed behavior, active inference makes associative and reinforcement learning (RL) [76] seem less like a tournament, but more like how these modes can be contextualized within a hierarchical structure. With this in mind, active inference provides the means of enhancing traditional RL methods by furnishing an inherent balance of exploration and exploitation, along with a more flexible concept of reward. In [77], these properties were demonstrated algorithmically in terms of a free energy model of the expected future. By how active inference frames an implicit scheme for adaptive control, adopting a hierarchical structure can be widely adaptable (for instance, as seen as layered robotic architectures). In particular, *hierarchical reinforcement learning* permits an agent, via a hierarchy, to aggregate actions into applicable reusable subroutines or skills [73]. *Motivated control* [72] can be seen as depending on hidden states of the world, and activation sequences (or policies) specifiable objectively; that is, in terms of beliefs about states of the world - in contrast to value judgements.[8] It is hierarchical active inference that produces coordinated control, and motivational aspects for joint optimization of actions sequences in respect of brain functions.[9] Often that can result in a control/motivation dichotomy, where a conflict between the two is a result of context, or vice-versa. For instance: "I like to swim to keep fit and healthy, but the sea is rough, and the strong current might drag me dangerously out to sea".

8.3.2 Tensor Networks and Control of Active Inference

Having seen how learning (of several types), goal-directed behavior, and overall motivation, are collectively factored into a hierarchical framework of active inference, let us return to the framework of Sect. 8.1.4 to see how the dynamics of these actions can be explicitly engineered for ANNs, applicable to ML and DL. Firstly, we observe that many standard TN models (such as MERAs) assume boundary conditions asymptotically far-reaching, in terms of numbers of lowest-scale operators, from the region of the network that is of interest. Identifying such asymptotic boundary conditions yields a cyclic system. Theorem 8.1.1, along with its corollaries, provides a natural, formal means of classifying systems by their control architectures. In view of the hierarchical architectures discussed above for

---

coding and neuroplasticity minimize a common cost function with modulated plasticity by delay, perform active inference. Such costs involving principles of thermodynamics arising from the action of pyramidal neurons, have been studied in [75]. As pointed out in [74], the accuracy of inference and decisions depend upon prior beliefs implicit in NNs. Clearly, synaptic plasticity associated with false priors results in less than optimal inferences and decisions for dealing with the engaged task.

[8] This is exemplified in [72] in terms of 'The Waiting Game'. The competition and gambits therein are seen at levels of policies, thereafter activating, or deferring, some decisive "go!" action. Hierarchical active inference, at least for learning, appears to run parallel to how schemes of nested MBs represent scales, or degrees of self-organization for systems in general [78].

[9] For multiple drives, policies, etc. this is analyzed in terms of the brain's dorsolateral areas (control) and ventromedial areas (motivation, or value) in [72] (cf. [79]).

associative and adaptive learning, for instance, let us note that in relationship, there are two salient characteristics for distinguishing systems with different architectures:

- The *hierarchical depth*, which indicates the number of "virtual machine" layers [80] that the architecture supports. The interfaces between these layers implement coarse-graining, removing from the higher-level representation all dimensions, and hence all of the information, which is contracted out of the lower-level operators.
- *The number and location of contractions that yield unitary operators*, and hence build in entanglement between lower-level operators. The natural limit is a MERA, in which every pair of lower-level operators is entangled at every hierarchical level [81].

Following these characteristics, the control-flow architecture specifies the structure of the "layout" of distinguishable sectors on $\mathscr{B}$, and hence of detectable features/objects in the environment. Locality on $\mathscr{B}$ requires a hierarchical TN, and detectable entanglement requires a MERA-like TN. Locality is required for detectable features/objects to appear to have components with nested decompositions. Any QRF for geometric space, and hence for spacetime, must be hierarchical, and hence must be a MERA if entanglement in space is to be detected. A MERA is required, in particular, if the use of coherence between spatially-separated systems as a computational or communication resource is detectable.

  Below, we will illustrate a hierarchically structured classification of systems by hierarchical level by firstly exhibiting a ten-step cyclic TN as shown below:

$$\overset{\frown}{A}\!-\!B\!-\!\cdots\!-\!J \qquad\qquad (8.12)$$

Then we extend Diagram (8.12) to a hierarchy as shown in Diagram (8.13) below:

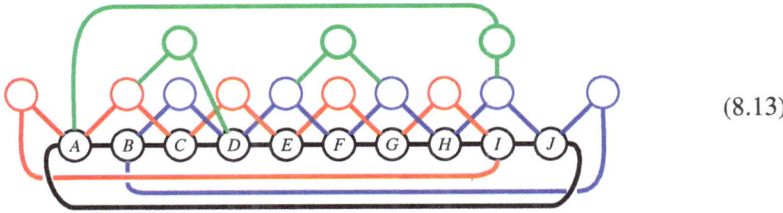

$$(8.13)$$

In (8.13), the red, blue, and green colors indicate distinct hierarchical layers of tensor contractions. There are several steps presented in [31] towards implementing control flow for hierarchical active inference which we will proceed to itemize (we kindly direct the reader with programming interests to [31, II–III]):

(1) Various tabulated Datasets of [31, II B Table 1] with specified control cycles instrument the training of *artificial neural networks* (ANNs) adapted to execute the corresponding TNs, which hence establish the latter as the desired representations of hierarchical active inference.

(2) The data of (1) is used to exemplify a corresponding tabulation of state transitions [31, II Fig. 1].

(3) Feedforward architecture is created to learn the control cycles of (1). Timesteps in the executed control flow are labeled accordingly: T (input) and $T + 1$ (output) [31, II Fig. 2].

(4) Architecture of the Bidirectional Associative Memory is exhibited in [31, II Fig. 3] with connections for the first (T - input) and last ($T + 1$ - output) nodes.

TNs have a close association with the TQNNs mentioned in Chap. 7, and hence with DNNs. The physical scalar products, seen as evolution matrix elements, induce a spanning in terms of spin network states. The TNs are, in practice, also classifiers for TQNNs, and the control aspect for active inference as we have outlined leads to connections with gauge field theory, and related differential geometric spacetime concepts (such as geometric, stochastic Ricci flow) that raise the theory significantly; we refer the interested reader to [31, III].

8.3.3   Control Flow and Biological Systems

Potential connections between TNs, ML and DL prevail widely in current computational biology (see for instance [55, 82] where the latter uses DL to study the 3-dimensional structure of proteins). It will be worth summarizing the key main ML architectures consisting of [82, Figure 1]:

(1) *Supervised learning*: Algorithms learn by comparing predicted and actual values.

(2) *Unsupervised learning*: Algorithms learn without labeled data.

(3) *Reinforcement learning*: An agent learns (via algorithms) by interaction with an environment.

and those of the DL architecture (in terms of NNs):

(1) *Convolutional NNs*: Perform inference on data with local features. Learn shift-invariant filters.

(2) *Recurrent NNs*: Perform inference on temporal data. Learn temporal relations via recurrent structure.

(3) *Transformer*: Perform inference on sequential data. Learn context based correlations via attention mechanism.

(4) *Autoencoder* (AE): Embed high-dimensional data. Learn low-dimensional embedding of data.

Note that these are, *a priori*, distributed systems for which the TQNN is, in principle, a carrier for the component functions of ML and DL in the control flow of active inference [5, 31]. Methods of researching deep temporal modes, as relevant, for instance, to working/episodal/procedural memory via NNs, commence by an agent minimizing VFE via sensory inputs and variational density to approximate posterior distributions inferring action as directed by the agent's GM; this is one key mechanism to accomplish 'understanding' in ML and DL [83, 84]. There are other explicit programmable methods for control flow in distributed ML and DL (see for instance, [85, 86]).

Another major source of applications involves biological systems modeled on scale-free architectures, particularly those which necessitate smooth transitions from quantum-like, to classical-like behavior. This is especially the case for metabolic, signal-transduction, and gene-regulatory pathways that are entirely classical, even though many of the intrinsic processes involve electron-transfer, or other mechanisms, that necessitate a quantum-like description [62, 63]. In principle, free-energy budget considerations account for why both prokaryotic and eukaryotic cells employ quantum coherence as a computational resource [87] (cf. [88]). The latter is also evidenced for large-scale networks involving longer-range entanglement in mammalian brains [89]. Whereas for prokaryotes, the primary tasks of control flow are for adapting metabolism to available resources, via metabolite-driven gene regulation [90], and initiating DNA replication and cell division when conditions are favorable.

These examples support the claim that control flow models are essential for supporting the likelihood of quantum computation in biological systems. Hierarchical TNs that include unitary components, for instance, MERA-type models as described earlier, provide this capability. Consequently, we can expect to find shallow hierarchies of effectively classical control transitions in these organisms. Eukaryotes, however, are characterized by both intracellular compartmentalization, and morphological degrees of freedom at the whole-cell scale. Significantly, instrumentation of the FEP will induce "neuromorphic" morphologies. These are morphologies that segregate inputs from outputs and enable a 'fan-in/fan-out' computational architecture in any systems with morphological degrees of freedom, as was shown in [91]. Such systems can be expected to have deep control hierarchies at the cellular level, with hierarchical structure correlating with morphological structure in morphologically-complex cells such as neurons represented by the mechanism of QRFs, as well as multicellular assemblages at all scales [92]. These distinctions correlate with the orders-of-magnitude increase in classical computational power (estimated from a total metabolic energy budget) as a function of cell-surface area in eukaryotes as compared to prokaryotes [87] (illustrated in [31, Fig. 6]).

8.3.4   Stigmergic Memory and Context-Sensitivity

Firstly, let us recall the concept of *stigmergy* as an emergent mechanism of complex systems, essential for self-organization entailing coordination within an environment between agents and actions, wherein the trace of an agent's action on some medium, stimulates the performance of a subsequent action by that, or other agents (see for instance [93]). Next, let us return to eukaryotes. For the best part, these life forms, besides managing metabolism and replication, implement an active exploration of the environment, communicating with other systems and, as is crucial for cognition, the writing and reading of stigmergic memories. Because of these properties, we can expect such systems to implement QRFs for spacetime, and for specific kinds of objects, for instance, conspecifics, as well suitable substrates for recording stigmergic memories. Such QRFs rely on symmetries, and hence on redundancy of encoded (or encodable) information; they depend, in other words, on the availability of error-correcting codes [87, 88]. The implementation of spacetime as a QECC by TNs has been extensively studied by physicists as noted above; see [94] for review and [29] for a detailed analysis using the present formalism (see also Chap. 7). The use of spacetime as an error-correcting code by organisms, that is via the implementation of translational and rotational invariance of objects by dorsal visual processing in mammals [95, 96], is well-understood phenomenologically, but the details of the actual neural implementation remain to be further researched. Further, in view of the outline of TNs in respect of ML and DL in Sect. 8.3.3, the report of [97] demonstrates how stimergic memory can be incorporated, and hence operative within the architectural design of a Recurrent NN, the latter seen as an agent of DL, for instance.

Both context-sensitivity, and the occurrence of context effects due to non-commutativity of QRFs influencing control networks, are expected to increase in proportion to their complexity and hierarchical depth. 'Bowtie' networks with high 'fan-in/fan-out' to/from multi-functional proteins, or second messengers such as Ca^{2+}, are increasingly ubiquitous in higher eukaryotic cells [98]. Such networks have the general form of the CCCD as depicted in Chap. 3, and as introduced elsewhere. Frequently, such networks evolve via compression of information (for instance towards share second messengers, as in $[Ca^{2+}]$-based interactions [99, 100]) as an efficiency-increasing mechanism. However, 'bowties' can introduce semantic ambiguities that must be resolved by establishing a context. Each incoming signal has its own governing semantics, but the relevant context can depend on boundary conditions which can be exceedingly difficult, if not impossible, to predetermine (see for instance [101, 102] for general surveys of the history, and the semantic depth of this problem). As pointed out in [42], a context change $x \mapsto y$ is *semantically problematic* if, for a fixed set $\{o_i\}$ of observations, the conditional probability distributions $P(o_i|x)$ and $P(o_i|y)$ are well defined, but the joint distribution $P(o_i|x \vee y)$ is not [103]. This occurs whenever the QRFs for x and y do not commute; it is an issue which can serve as a convenient benchmark for distinguishing signal transduction networks that incorporate co-deployable, versus non-co-deployable QRFs [40] (in particular [40, Theorem 7.1]; see also [41]).

The discussion of Chap. 5 relating to contextuality, and "quantum-like" effects, predicts that a context-switching problem affects DL using VAEs [67], as for instance in the application to antimicrobial peptides in [104]. Much of our account in Chap. 5 applies equally well to 'language' in the broadest sense. While the latter has thus far been applied to cells, tissues, and even non-vertebrates in a mostly metaphorical way, we can speculate that linguistic approaches to understanding the interplay between context dependence and semantic ambiguity, may eventually prove to be of paramount importance throughout the biological sciences. Immune cells (such as T cells) are, for example, 'programmed' or 'trained' by their progenitor cells to respond to local cellular signals and ambient conditions in particular ways. Unexpected changes in context expectedly induce dysfunctional (at the organism scale) responses, which may include chronic disorders [105]. They are manifestly the outcomes of discrepancies between the "actual" semantics of incoming signals, and the semantic structure to be activated by the immune system's "language." This suggestion of possible "linguistic contextuality" appears consonant with the hypothesis of [106]; namely, that the immune system is a cognitive (living) system implementing its exclusive system of language-grammar, which may be prone to disorders of communication analogous to those discussed in [107]. Certain microbiological systems have revealed effects which exhibit contextuality. For instance, discrepancies in experimentally derived classical probabilities arising from lactose-glucose interference signaling in *E. Coli* can only be explained in terms of non-classical probabilities [108] (these are the typical "quantum-like" effects according to [109], where we note that the formal structure is exactly that as mandated by quantum theory).

References

1. Friston, K.J.; Da Costa, L.; Sakthivadivel, D.A.R.; Heins, C.; Pavliotis, G.A.; Ramstead, M.J.; Parr, T. Path integrals, particular kinds, and strange things. Phys. Life Rev. 2023, 47, 35–62.

2. Friston, K. J., Rigoli, F., Ognibene, D., Mathys, C., Fitzgerald, T. and Pezzulo, G. Active inference and epistemic value. *Cognitive Neuroscience* 6 (2015), 187–214.

3. Friston, K. J., Lin, M., Frith, C. D., Pezzulo, G., Hobson, J. A. and Ondobaka, S. Active inference, curiosity and insight. *Neural Comput.* 29(10) (2017), 2633–2683.

4. Friston, K. Life as we know it. J. R. Soc. Interface 2013, 10, 20130475.

5. Fields, C.; Fabrocini, Friston, K.; Glazebrook, J.F.; Hazan, H.; Levin, L.; Marcianò, A. Control flow in active inference systems, Part I: Formulations of classical and quantum active inference. IEEE Trans. Mol. Biol. Multi-Scale Commun. 2023, 9, 235–245.

6. Schwartenbeck, P., Passecker, J., Hauser, T., FitzGerald, T. H. B., Kronbichler, M. and Friston, K. J. Computational mechanisms of curiosity and goal-directed explorarion. *bioRxiv* (2018), 411272.

7. Friston, K.; FitzGerald, T.; Rigoli, F.; Schwartenbeck, P.; Pezzulo, G. Active inference: A process theory. Neural Comput. 2017, 29, 1–49.

8. Friston, K. FitzGerald, T., Rigoli, F., Schwartenbeck, P., O'Doherty. J. and Pezzulo, G. Active inference and learning. *Neuroscience and Biobehavioral Reviews* 68 (2016), 862–879.

9. Çatal, O., Nauta, J., Verbelen, T., Simoens, P. and Dehoedt, B. Bayesian policy selection using active inference. In, *Proceedings of the workshop on Structures & Priors in Reinforcement Learning* at ICLR 2019.

10. Fields, C. and Glazebrook, J. F. Nash equilibria and undecidability in generic physical interactions - A free energy perspective. *Games* 15(30) (2024).

11. Fields, C.; Friston, K.J.; Glazebrook J.F.; Levin M. A free energy principle for generic quantum systems. Prog. Biophys. Mol. Biol. 2022, 173, 36–59.

12. Friston, K.J. A free energy principle for a particular physics. arXiv 2019, arXiv:1906.10184.

13. Attias, H. Planning by probabilistic inference. *Proc. of the 9th Int. Workshop on Artificial Intelligence and Statistics* in Proc. Machine Learning Res. R4 (2003), 9–16.

14. Botvinick, M. and Toussaint, M. Planning as inference. *Trends Cogn. Sci.* 16(10) (2012), 485–488.

15. Lanillos, P., Mio, C., Pezzato, C., et al. Active inference in robotics and artificial agents: Survey and challenges. (2021) Preprint arXiv:2112.01871.

16. Kirchoff, M., Parr, T., Palacios, E., Friston, K. and Kilverstein, J. The Markov blankets of life: autonomy, active inference and the free enrgy principle. *J. R. Soc. Interface* 15 20170792.

17. Hipólito, I., Ramstead, M. J. D., Convertino, L., Bhat, A., Friston, K. and Parr, T. Markov blankets in the brain. *Neurosci. Biobehv. Rev.* 125 (2021), 88–97.

18. Conway, J.; Kochen, S. The free will theorem. Foundations of Physics 2006; 36:1441–1473.

19. Conway, J.; Kochen, S. The strong free will theorem. Notices of the AMS 2009; 56:226–232.

20. Milnor, J. *Games Against Nature.* RAND Corp.: Santa Monica, CA, USA, 1951.

21. Ekert, A. K., Huttner, B., Palmer, G. M. and Peres, A. Eavesdropping on quantum cryptographical systems. *Phys. Rev. A* 50 (1994), 1047–1056.

22. Addazi, A.; Chen, P.; Fabrocini, F.; Fields, C.; Greco, E.; Lutti, M.; Marcianò, A.; Pasechnik, R. Generalized holographic principle, gauge invariance and the emergence of gravity à la Wilczek. Front. Astron. Space Sci. 2021; 8: 563450.

23. Fields, C. Glazebrook, J. F.; Marcianò, A. The physical meaning of the Holographic Principle. Quanta 2022; 11:72–96.

24. Fields C., Marcianò, A. Holographic screens are classical information channels. *Quant. Rep.* 2 (2019), 326–336.

25. Fields, C.; Glazebrook, J.F.; Marcianò, A. Reference frame induced symmetry breaking on holographic screens. Symmetry 2021, 13, 408.

26. Fields, C.; Glazebrook, J.F. Representing measurement as a thermodynamic symmetry breaking. Symmetry 2020, 12, 810.

27. Tipler, F. Quantum nonlocality does not exist. *Proc. Natl. Acad. Sci. USA* 111 (2014), 11281–11286.

28. Hooft, G.T. Deterministic quantum mechanics: The mathematical equations. *Front. Phys.* 8 (2020), 253.

29. Fields, C., Glazebrook, J. F. and Marcianò, A. Communication protocols and QECCs from the perspective of TQFT, Part I: Constructing LOCC protocols and QECCs from TQFTs. Fortschritte der Physik 2024; 72: 202400049

30. Wigner, E.P. Remarks on the mind-body question. In (I. J. Good, ed.) *The Scientist Speculates* pp. 284–302. Heinemann: London, UK, 1961.

31. Fields, C., Fabrocini, F., Friston, K., Glazebrook, J. F., Hazan, H., Levin, M. and Marcianò, A. (2023) Control flow in active inference systems, Part II: Tensor networks as general models of control flow. IEEE Transactions on Molecular, Biological, and Multi-Scale Communications 9: 246–256

32. Adams, J. F. *Lectures on Lie Groups.* Univ. Chicago Press, Chicago IL, 1983.

33. Hermann, R. *Lie Groups for Physicists.* W. A. Benjamin, New York, 1966.

34. Carette, T., Glowacki, J. and Loveridge, L. Operational quantum reference frames. arXiv:2303.14002 (2023).

35. Palmer, M. C., Girelli, F. and Bartlett, S. D. Changing quantum reference frames. *Phys. Rev. A* 89 (2014), 052121.

36. Fields, C., Glazebrook, J. F., Marcianò, A., and Zappala, E. (2025) ER = EPR is an operational theorem. Physics Letters B 860, 139150.

37. de la Hamette, A.-C. and Galley, T. D. Quantum reference frames for general symmetry groups. *Quantum* 4 (2020), 367.

38. Seifert, U. Stochastic thermodynamics, fluctuation theorems and molecular machines. *Rep. Prog. Phys.* 75 (2012), 126001.

39. Chitambar, E., Leung, D., Mančinska, L., Ozols. M. and Winter, A. Everything you always wanted to know about LOCC (but were afraid to ask). Commun. Math. Phys. 2014; 328: 303–326.

40. Fields, C.; Glazebrook, J. F. Information flow in context-dependent hierarchical Bayesian inference. *J. Expt. Theor. Artif. Intell.* 34 2022, 111–142.

41. Fields, C.; Glazebrook, J. F. Separability, contextuality, and the quantum Frame Problem. International Journal of Theoretical Physics 2023; 62: 159.

42. Fields, C., Glazebrook, J. F. and Levin, M. Minimal physicalism as a scale-free substrate for cognition and consciousness. *Neuroscience of Consciousness* 7(2) 2021, niab013.

43. Bateson, G. *Steps to an Ecology of Mind: Collected Essays in Anthropology, Psychiatry, Evolution, and Epistemology*. Jason Aronson, Northvale, NJ, USA, 1972.

44. Roederer, J. (2005). *Information and its Role in Nature*. Heidelberg: Springer.

45. Fields, C.; Levin, M. How do living systems create meaning? Philosophies 2020; 5: 36.

46. Fields, C. and Levin, M. (2025). Thoughts and thinkers: On the complementarity between objects and processes. Physics of Life Reviews 52: 256–273

47. Fields, C. and Levin, M. (2023) Regulative development as a model for origin of life and artificial life studies. BioSystems 229: 104927

48. Fields, C. and Levin, M. (2025) Life, its origin, and its distribution: a perspective from the Conway-Kochen Theorem and the Free Energy Principle. Communicative & Integrative Biology 18: 2466017.

49. Sengupta, R., Adhikary, S., Oseledets, I. and Biamonte, J. Tensor networks in machine learning. *Euro. Math. Soc. Mag.* 126 (2022), 4–12. Preprint arxiv:2207.02851, 2022.

50. Jahroni, S. S. and Orús, R. Variational tensor neural networks for deep learning. *Scientific Reports - Nature Portfolio* 14 (2024), 19017.

51. Qi, X. L. Does gravity come from quantum information? *Nature Physics* 14 (2018), 984–987.

52. Guyon, I. and Elisseeff, A. An introdution to variable and feature selection. *J. of Machine Learning Res.* 3 (2003), 1157–1182.

53. Guyon, I., Aliferis, C. and Elisseeff, A. Causal feature selection. In (H. Liu and H. Motoda, eds.) *Computational Methods of Feature Selection*, Ch 4. pp. 63–85. CRC Press, Boca Raton, FL, 2007.

54. Pellet, J.-P. and Elisseeff, A. Using Markov blankets for causal structure learning. *J. Machine Learning Research* 9 (2008), 1295–1342.

55. LeCun, Y., Bengio, Y. and Hinton, G. Deep Learning. *Nature* 521 (2015), 436–444.

56. Bao, N., Cao, C., Carroll, S. M. and Chatwin-Davies, A. De Sitter space as a tensor network: cosmic no-hair, complementarity, and complexity. *Phys. Rev. D* 96 (2017), 123536.

57. Hu, Q. and Vidal, G. Spacetime symmetries and conformal data in the continuous Multiscale Entanglement Renormalization Ansatz. *Phys. Rev. Lett.* 119 (2017), 010603.

58. Chandra, A. R., de Boer, J., Flory, M., Heller, M. P., Hörtner, S. and Rolph, A. 2021 Spacetime as a quantum circuit. *J. High Energy Phys.* (2021), 207.

59. Pezzato, C., Corbato, C. H., Bonhof, S. and Wisse, M. Active inference and behavior trees for reactive action planning and execution in robotics. *IEEE Trans. Robotics* 39(2) (2023), 1050–1069.

60. Proietti, R., Pezzulo, G. and Tessari, A. An active model of hierarchical action, understanding and learning. *Physics of Life Reviews* 46 (2020), 92–118.

61. Chubukov V., Gerosa, L., Kochanowski, K., and Sauer, U. Coordination of microbial metabolism. *Nat. Rev. Microbiol.* 12 (2014), 327–340.

62. Zweir, M. C. and Chong, L. T. Reaching biological timescales with all-atom molecular dynamics simulations. *Curr. Opin. Pharmacol.* 10 (2010), 745–752.

63. Groenhof, G. Introduction to QM/MM simulations. *Methods Mol. Biol.* 924 (2013), 43–66.

64. de Vries, B. and Friston, K. J. A factor graph description of deep temporal active inference. *Front. Comput. Neurosci.* 11 (95) (2017), 16 pp.

65. Dauwels, . On variational message passing on factor graphs. *2007 IEEE International Symposium on Information Theory* pp. 2546–2550. Nice, France, 2007.

66. Winn, J. and Bishop, C. M. Variational message passing. *J. Mach. Learn. Res.* 6 (2005), 661–694.

67. Kingma, D. P and Welling, M. An Introduction to variational autoencoders. *Foundations and Trends in Machine Learning* 12(4) (2019), 307–392.

68. Levin, M. Darwin's agential materials: Evolutionary implications of multiscale competency in developmental biology. *Cell. Mol. Life Sci.* 80 (2023), 142.

69. Clawson, W. and Levin, M. Endless forms most beautiful 2.0: Teleonomy and the bioengineering of chimaeric and synthetic organisms. *Biol. J. Linnean Soc.* 2022, 2022, blac073.

70. Dayan, P., Hinton, G. E., Neal, R. M. and R. S. Zemel. The Helmholtz machine. *Neural Computation* 7 (1995), 1022–1037.

71. Feldman, H. and Friston, K. (2010). Attention, uncertainty, and free energy. *Front. Hum. Neurosci.* 4 (2010), 215.

72. Pezzulo, G., Rigoli, F. and Friston, K. J. Hierarchical active inference: A theory of motivated control. *Trends in Cognitive Sciences* 22(4), (2018), 294–306.

73. Botvinick, M. M., Niv, Y. and Barto, A. C. Hierarchially organized behavior and its neural foundations: a reinforcement-learning perspective. *Cognition* 113(3) (2009), 262–280.

74. Isomura, T., Shimazaki, H. and Friston, K. J. Canonical neural networks perform active inference. *Commun. Biology* 5(55) (2022), 15 pp.

75. Georgiev, D. D., Kolev, S. K., Cohen, E. and Glazebrook, J. F. Computational capacity of pyramidal neurons in the cerebral cortex. *Brain Research* 1748 (2020), 147069.

76. Sutton, R. S. and Barto, A. G. *Reinforcement Learning: An Introduction.* MIT Press, Cambridge MA, 1998.

77. Tschantz, A., Millidge, B., Seth, A. K. and Buckley, C. L. Reinforcement learning through active inference, Workshop paper at *Bridging AI and Cognitive SCience.* International Conference on Learning Representations (ICLR) (2020), 14 pp.

78. Ramstead, M. J. D., Badcock, P. D. and Friston, K. J. Answering Schrödinger's question. A free-energy formulation. *Physics of Life Reviews* 24 (2018), 1–16.

79. Wright, J. and Bourke, P. Markov blankets and mirror symmetries: Free energy minimization and mesocortical anatomy. *Entropy* 26 (2024), 287.

80. Smith, J. E. and Nair, R. The architecture of virtual machines. *IEEE Computer* 38(5) (2005), 32–38.

81. Orús, R. (2019). Tensor networks for complex quantum systems. *Nat. Rev. Phys.* 1, 538–550.

82. Sapoval, N., Aghazadeh, A., Nute, M. G., et al. Current progress and open challenges for applying deep learning across the biosciences. *Nature Commun.* 13 (2022), 1728.

83. Parr, T. and Pezzulo, G. Understanding, explanation, and active inference. *Front. Sys. Neuro.* 15 (2021), 772641.

84. Ueltzhöffer, K. Deep active inference. *Biol. Cybern.* 112(6) (2018), 547–573.

85. Tiwari, T., Gururangan, S., Guo, C., et al. Information flow control in machine learning through modular model architecture. arXiv:2306.03235v2 [cs.LG] (2024).

86. Yu, Y. Abadi, M., Barham, P. et al. Dynamic control flow in large-scale machine learning. *EuroSys 2018, 13th EuroSys Conference.* April 23-28, 2018, Porto, Portugal. ACM, New York, NY.

87. Fields, C. and Levin, M. Metabolic limits on classical information processing by biological cells. *BioSystems* 209 (2021) art. 104513.

88. Fields, C.; Levin, M. Multiscale memory and bioelectric error correction in the cytoplasm–cytoskeleton-membrane system. WIREs Syst Biol Med 2017, e1410.

89. Kerskens, C. M. and Pérez, D. L. Experimental indications of non-classical brain functions. *J. Phys. Commun.* 6 (2022), art. 105001.

90. D. Ledezma-Tejeida, E. Schastnaya, and U. Sauer, Metabolism as a signal generator in bacteria," *Curr. Opin. Syst. Biol.*, 28, 100404, 2021.

91. Fields, C., Friston, K., Glazebrook, J. F., Levin, M. and Marcianò, A. (2022) The free energy principle induces neuromorphic development. Neuromorphic Computing and Engineering 2: 042002

92. Fields, C., Glazebrook, J. F. and Levin, M. Neurons as hierarchies of quantum reference frames. *Biosystems* 219 (2022), art. 104714.

93. Heylighen, F. Stigmergy as a universal coordination mechanism I: Definition and components. Cognitive Systems Research 2016; 38: 4-13.

94. Bain, J. Spacetime as a quantum error correcting code? Stud. Hist. Phil. Sci. B 2020; 71: 26-36.

95. Flombaum, J. I., Scholl, B. J. and Santos, L. R. (2008). Spatiotemporal priority as a fundamental principle of object persistence. In (Hood, B. Santos, L. eds.) *The Origins of Object Knowledge* pp. 135–164. Oxford University Press UK, 2008.

96. Fields, C. Trajectory recognition as the basis for object individuation: A functional model of object file instantiation and object token encoding. *Front. Psychol.: Percept. Sci.* 2 (2012), 49.

97. Galatolo, F. A., Cimino, M. G. C. A. and Vaglini, G. Using stigmergy as a computational memory in the design of recurrent neural networks. *Proc. 8th Int. Conf. Pattern Recognition Applications and Methods*, SCITEPRESS - Science and Technology Publications, 2019. arXiv:1903.01341v1 [cs-NE]

98. Niss, K., Gomez-Casado, C., Hjaltelin, J. X., Joeris, T., Agace, W. W., Belling, K. G. and Brunak, S. Complete topological mapping of a cellular protein interactome reveals bow-tie motifs as ubiquitous connectors of protein complexes. *Cell Rep.* 31 (2020), art. 107763.

99. Carafoli, E. and Krebs, J. Why calcium? How calcium became the best communicator. *J. Biol Chem* 40 (2016), 20849–20857.

100. Polouliakh, N., Nock, R., Nielsen, F. and Kitano, H. G-protein coupled receptor siganling architecture of mammalian immune cells. *PLoS ONE* 4 (2009), art. e4189.

101. Friedlander, T., Mayo, A. E., Tlusty, T. and Alon, U. Evolution of bow-tie architectures in biology. *PLOS Computational Biology* 11 (2015), art. e1004055.

102. Boniolo, G., D'Agostino, M., Piazza, M. and Pulcini, G. Molecular biology meets logic: Context-senstivity in focus. *Found. Science* 28(1) (2021), 307–325.

103. Kochen, S., Specker, E. P. (1967). The problem of hidden variables in quantum mechanics. Journal of Mathematics and Mechanics 17, 59–87.

104. Dean, S. N. and Walper, S. A. Variational autoenecoder for generation of antimicrobial peptides. *ACS Omega* 5 (2020), 20746–20754.

105. Editorial Focus. A matter of context. *Nature Immunol.* 20 (2019), 769.

106. Atlan, H. and Cohen, I. R. Immune information, self-organiztion and meaning. *Int. Immunology* 10 (1999), 711–717.

107. Wang, D., Sadrzadeh, Abramsky, S. and Cervantes, V. H. On the quantum-like contextuality of ambiguous phrases. *Proceedings of the 2021 Workshop on Semantic Spaces at the Intersection of NLP, Physics and Cognitive Science*, pp. 42–52. Association for Computational Linguistics 2021.

108. Basieva, I., Khrennikov, A., Ohya, M. and Yamato, O. Quantum-like interference effect in gene expression: glucose-lactose destructive interference. *Syst Synth Biol* 5 (2011), 59–68.

109. Khrennikov A (2015) Quantum-like modeling of cognition. Front. Phys. 3:77. https://doi.org/10.3389/fphy.2015.00077

How Much of Biology Is "Just Physics"?　9

We began this book in Chap. 1 by tracing the gradual development, from Clausius and Boltzmann in the 19th century to current developments in the 21st, of the idea that physical interactions can all, without exception, be viewed as implementing information transfer. Provided the interaction is finite in space, time, and energy, the information transferred is finite. The symmetry of the Hamiltonian—$H_{SE} = H_{ES}$, Newton's Third Law, or the discrete nature of finite holographic screens—guarantees that the information transfer is also symmetrical. All separable, multi-component physical systems are, therefore, distributed information-flow systems. Indeed they are *communication* systems. In the language of the FEP, they are self-organizing active-inference agents.

While self-organization has been a theme in physics since the early 1970s, it has been a theme—arguably *the* theme—in biology at least since Darwin. Arguments still abound in both the academic and the popular press that biology is "beyond physics" or at least beyond current physics. With the universality of the FEP in mind, however, we must ask whether "life" can sensibly be defined—except ostensively as "life as we know it" [1]—in a way that leaves living systems a special case of physical systems [2]. We must also, then, ask the reverse question: if *all* physical systems are alive in at least some limited sense, or even if *most* physical systems are alive in at least some limited sense, what impact does this have on physics? Setting aside the obvious moral questions, what impact would this have on the methodology of physics? Adopting the theoretical methodology of the FEP gives this question a special urgency: Is the FEP, in fact, a theoretical methodology for a universalized physics of living systems, one that allows us to describe all physical systems as self-organizing, mutually communicating, active inference agents? Our conclusions throughout this book suggest that it is.

That all physical systems are *cognitive* in at least some limited sense was already suggested by Turing [3]. The development here makes this even more evident. Any finite com-

C. Fields and J. Glazebrook, *Distributed Information and Computation in Generic Quantum Systems*, Synthesis Lectures on Engineering, Science, and Technology, https://doi.org/10.1007/978-3-031-97263-8_9

bination of logical gates—classical or quantum—can be implemented by a finite CCCD [4]; hence any Turing machine, classical or quantum, can be implemented by a finite CCCD. Universality in this computational sense underlies the ability of CCCDs to represent any finite QRF, and hence the mapping from CCCDs to TQFTs [5]. The FEP adds to the identification of physical dynamics with computation—an identification already implicit in Fig. 1.2—that *persistent* dynamics can always be described as Bayesian satisficing.

What physical systems can compute, even up to the cosmological scale, is ultimately limited by thermodynamics [6]. Obtaining thermodynamic resources, including an environment in which to dissipate waste heat, is therefore a prime concern of any autonomous computing system. This is obvious for organisms, but has largely been neglected for generic systems [7]. Even if neural systems can be shown to implement quantum computations [8], we can expect a primary goal of such computations to be the maintenance of adequate thermodynamic resources. The fundamental lesson of the FEP is that persistence is not free—indeed, unitary evolution or even classical entropic flow destroys it—so any persistent system must actively maintain itself. Computational processes—cognition—ar always in service of their own preservation.

Having said this, we must add a *proviso* based on Stephen Hawking's contention [9] that physics is incomplete. Hawking's claim was influenced by the nature of Gödel's incompleteness theorems as founded, to an extent, on the concept of 'self-referentiality'. According to [9], physical systems can be self-referencing, and thus are expected to be incomplete and/or inconsistent, as present-day physical theories are indeed widely deemed to be (cf. [10]). As an example, let us take Heisenberg's Uncertainty Principle as fundamental in quantum mechanics. As a physical principle, it is incomplete, and as a principle of indeterminacy, it states a form of quantum contextuality (these conclusions, along with others, are reviewed in [11, 5.1–5-2]).

While Einstein viewed incompleteness as a theological deficit, Bell wrote it off to lack of imagination, and Hawking and many others see it as a brake on human hubris, our development here shows that it is not human-specific but universal, and precisely identifies its source. A finite set of mutually-commuting operators acting on a proper subset of the qubits composing a holographic screen cannot determine even the dimension of the holographic screen, much less the dimension of the arbitrarily-large system on the other side [11, 12]. The incompleteness of physics, indeed the incompleteness of observational outcomes of any kind, is rooted in the physical conditions for maintaining separability. No agent—no persistent entity that maintains a well-defined individual state—can fully know its environment, and no such entity can fully know itself.

References

1. Mariscal, C; Doolittle, W. F. Life and life only: A radical alternative to life definitionism. Synthese 2020, 197, 2975–2989.

2. Fields, C. and Levin, M. (2025) Life, its origin, and its distribution: a perspective from the Conway-Kochen Theorem and the Free Energy Principle. Communicative & Integrative Biology 18: 2466017.

3. Turing, A. Computing machinery and intelligence. *Mind* 59, 433–460, 1950.

4. Fields, C. and Glazebrook, J. F. (2019a). A mosaic of Chu spaces and Channel Theory I: Category-theoretic concepts and tools. *Journal of Experimental and Theoretical Artificial Intelligence* 31(2), 177-213.

5. Fields, C., Glazebrook, J. F. and Marcianò, A. Sequential measurements, topological quantum field theories, and topological quantum neural networks. Fortschritte der Physik 2022; 70: 202200104.

6. Lloyd, S. Ultimate physical limits to computation. *Nature* 406 (2000), 1047–1054.

7. Ororbia, A., Friston, K. Mortal computation: A foundation for biomimetic intelligence. Preprint arxiv:2311.09589, 2023.

8. Swan, M., Dos Santos, R. P., Lebedey, M. and Witte, F. *Quantum computing fot the brain* (*Between Science and Economics* vol. 3). World Scientific Publ., London UK, 2022.

9. Hawking, S. Gödel and the End of Physics. Lecture at the *Dirac Centennial Celebration*. Centre for Mathematical Sciences, University of Cambridge, Cambridge UK. https://www.damtp.cam.ac.uk/events/strings02/dirac/hawking.html

10. Wheeler, J. A. The computer and the universe. *Int. J. Theor. Phys.* **1982**, *21*, 557–572.

11. Fields, C.; Glazebrook, J. F.; Levin, M. Principled limitations on self-representation for generic physical systems. *Entropy* **2024**, *26*, 194.

12. Fields, C.; Glazebrook, J. F. Separability, contextuality, and the quantum Frame Problem. International Journal of Theoretical Physics 2023; 62: 159.

Basic Category Theory Concepts

A

A.1   Objects and Arrows, Functors

Category Theory (CT) provides a conceptual tool-box adaptable to the descriptive methods of complex processes as these are encountered in computer science, (including artificial intelligence), physics, the life sciences, philosophy and logic, and many related fields. Since CT was introduced in [1], and formalized in [2], there is a range of more recent studies of the subject as can be found, for instance, in [3, 4]. Since a number of elements of CT have been used occasionally throughout the book, it befits us to guide the reader through some of the very basic concepts.

We recall that a *category* $\mathfrak{C}$ consists of collections of *objects*, and *arrows* (i.e. directed relations, or morphisms) between objects, satisfying two conditions: (1) arrows compose associatively, i.e. for objects A, B, C, D, if $f : A \rightarrow B$, $g : B \rightarrow C$, and $h : C \rightarrow D$, then $hgf : A \rightarrow D$, and (2) each object has an identity arrow $\mathrm{id}_A : A \rightarrow A$, where composition respects identities, and hom sets are pairwise distinct; for further details, see [3, 4]. There is a plethora of familiar mathematical instances involving such pairing of concepts. These include sets with functions, sets with relations, vector spaces with linear mappings, topological spaces with continuous mappings, groups with group homomorphisms, measure spaces with measurable functions, as the respective objects and arrows in each case.

A motivating example is that of a *Groupoid*: a 'small' category (where objects and arrows comprise sets) in which every arrow is invertible. A groupoid generalizes the 'group' concept in so far that the former can admit multiple 'identities' [5].

A *functor* $F : \mathfrak{C} \longrightarrow \mathfrak{D}$, between categories $\mathfrak{C}$ and $\mathfrak{D}$ is a mapping of objects to objects, and arrows to arrows, such that:

© The Editor(s) (if applicable) and The Author(s), under exclusive license to Springer
Nature Switzerland AG 2026
C. Fields and J. Glazebrook, *Distributed Information and Computation in Generic Quantum Systems*, Synthesis Lectures on Engineering, Science, and Technology,
https://doi.org/10.1007/978-3-031-97263-8

1. $F(f : A \longrightarrow B) = F(f) : F(A) \longrightarrow F(B)$,
2. $F(\mathrm{id}_A)) = \mathrm{id}_{F(A)}$,
3. $F(g \circ f) = F(g) \circ F(f)$.

(for further details and properties of functorial relations, see e.g. [3, 4]).

A.2   Cocone and Colimit

We have seen how the cocone and colimit concepts have featured in local-to-global distributed information flow in Channel Theory in the guise of a 'core'. Here we provide a descriptive account to enable the reader to get a feeling for these concepts (as treated more formally in e.g. [3, 4]). We will follow the account given in [6, 7].

First, we can specify the 'input data' for a colimit in terms of a *diagram* **D**, i.e. a collection of some objects in a category $\mathfrak{C}$, together with some arrows between them, as depicted by:

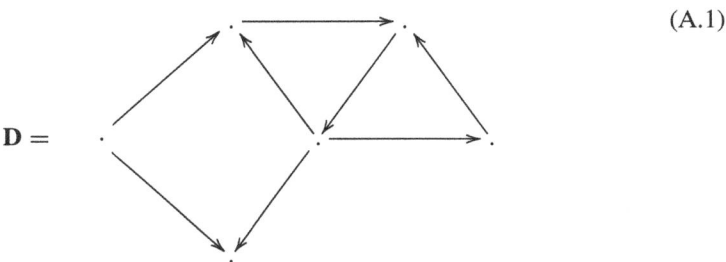

$$(A.1)$$

This diagram exhibits the *vertices* and *edges* of **D** to be the objects and morphisms of an arbitrary category. Next, we need 'functional controls' comprising a *cocone with base* **D** *and vertex as an object* **C** *in* $\mathfrak{C}$:

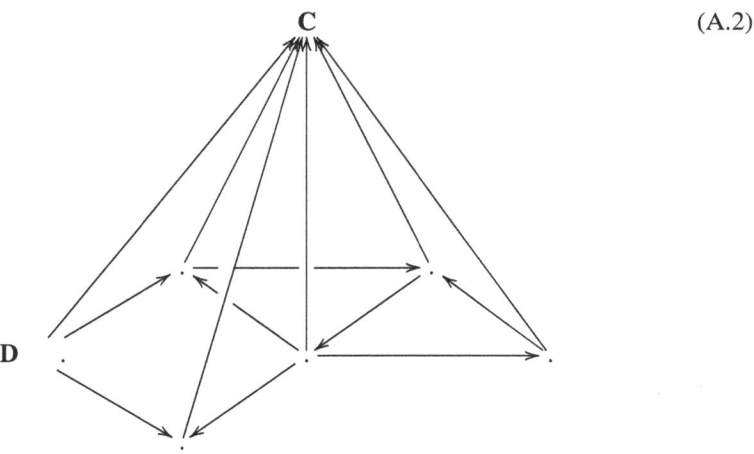

$$(A.2)$$

such that each of the triangular faces of the cocone is commutative. The 'output' from **D** will be an object colim(**D**) in our category $\mathfrak{C}$, as defined by a special *colimit cocone*, such that any cocone on **D** factors *uniquely* through the colimit cocone. Effectively, the commutativity condition on the cocone induces, in the colimit, an interaction of images from different parts of the diagram **D**. The uniqueness condition makes the colimit the optimal solution to this factorization problem.

Let

$$\blacklozenge = \text{colim}(\mathbf{D}) \qquad\qquad (A.3)$$

where the dotted arrows in the diagram below represent new morphisms which combine to make the colimit cocone:

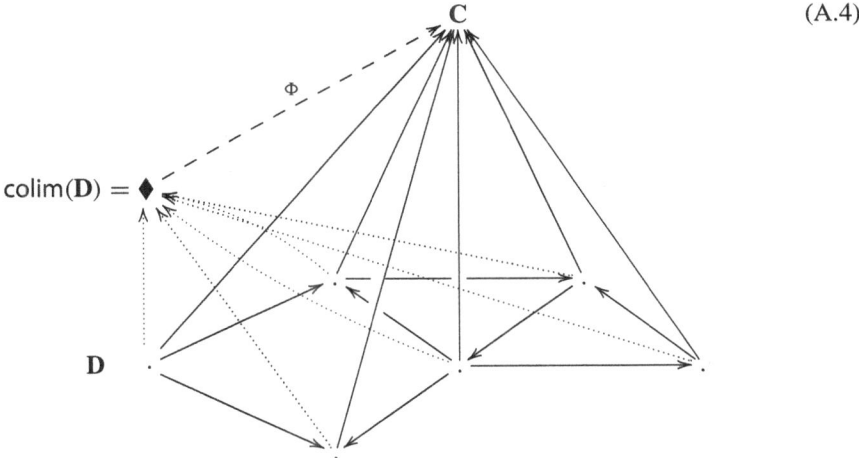

$$(A.4)$$

and for which the broken arrow Φ is constructed by requiring commutativity for all of the triangular faces of the combined diagram. Stripping away the former cocone results in a factorization of the cocone via the colimit:

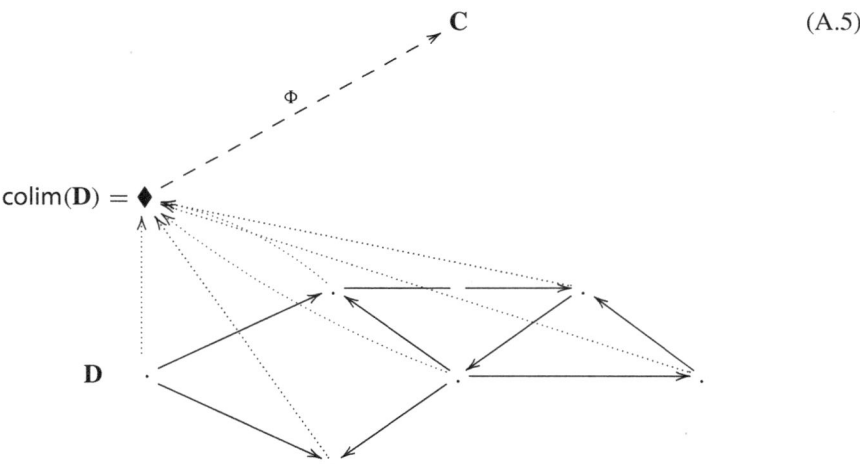

(A.5)

Intuitively, the object $\mathsf{colim}(\mathbf{D})$ is pieced together from the diagram $\mathbf{D}$ by means of the colimit cocone. From beyond $\mathbf{D}$, an arbitrary object $\mathbf{C}$ in $\mathfrak{C}$ 'sees' $\mathbf{D}$ as mediated through its colimit. This means that if $\mathbf{C}$ is poised to interact with all of $\mathbf{D}$, then it does so via $\mathsf{colim}(\mathbf{D})$.

The colimit cocone can be thought of as a kind of program: given any cocone on $\mathbf{D}$ with vertex $\mathbf{C}$, the output will be a morphism

$$\Phi : \mathsf{colim}(\mathbf{D}) \to \mathbf{C} \tag{A.6}$$

as constructed from the other data.[1]

Example A.2.1 Here is an analogy between colimits and distributed systems of information involving that of relaying an email message [7]. Suppose E denotes some email document. This is to be sent via a server S, which decomposes E into numerous parts E_i ($i \in \mathcal{I}$, an indexing set), and labels each part E_i, so it becomes E_i'. These labelled parts E_i' are then sent to a number of servers S_i, which then relay these messages as newly labelled messages E_i'' to a server S_C, for the receiver C. The server S_C then combines the E_i'' to produce the recovered message M_E at C. Breaking the message down, and routing it through the S_i appears arbitrary, but the system is designed such that M_E is independent of all choices made at each step of the process.

The colimit concept, as applicable to applications of category theory, is further discussed in [8]. It features in the study of Memory Evolutive Systems in [9], as well as prominently in our own work, for instance [10–13].

---

[1] The diagrams included in (A.1)–(A.5) are reproduced from [6, 7], with permission from R. Brown.

Chu Spaces

B

Chu spaces were first conceived by M. Barr and his student P. H. Chu as a categorical concept more general than that of topological spaces [14, 15]. Thanks to the work of V. Pratt and others, they found their way to multiple applications in theoretical computer science, and associated logical structures as applicable to a number of sub-fields including higher dimensional automata, concurrency, event space structures, and Petri nets [16–19] (see also [20]). Chu spaces are equivalent to the Barwise-Seligman classifiers/classifications of [21]. They have also featured in a number of other fields which include physical systems, contextuality, formal concept analysis and ontologies, as reviewed and exemplified in [10, 11, 22]. We will proceed to review some of the basic concepts of the theory.

B.1   The Basic Concepts of Chu Spaces

Definition B.1.1 Let $\mathbf{K}$ be a set with no structure assumed. A *Chu space* $\mathcal{A} = (A, r, X)$ *over* $\mathbf{K}$ consists of sets A and X, and a *satisfaction relation* (or *evaluation function*) $r : A \times X \longrightarrow \mathbf{K}$.

Usually, the sets A and X are referred to as *Objects* and *Attributes*, respectively, following which $\mathcal{A}$ comprises an object in the category of Chu spaces, denoted by **Chu(Set, K)**, or more simply as **Chu**. Other nomenclature can be used: for instance, a Chu space $\mathcal{A} = (A, r, X)$ over a set $\mathbf{K}$, called the *alphabet*, consists of a set A of *points* constituting the *carrier*, a set X of *states* constituting the *cocarrier*, and a function $r : A \times X \longrightarrow \mathbf{K}$, called the *matrix* [18] (see also Chu spaces as *types* and *processes* in [17]). The Tokens and Types of Clasifiers/Classifications of the isomorphic category of Channel Theory [21], are likewise, with ⊩ playing the role of the satisfaction relation r above.

© The Editor(s) (if applicable) and The Author(s), under exclusive license to Springer Nature Switzerland AG 2026
C. Fields and J. Glazebrook, *Distributed Information and Computation in Generic Quantum Systems*, Synthesis Lectures on Engineering, Science, and Technology, https://doi.org/10.1007/978-3-031-97263-8

Chu spaces are frequently amenable to matrix representations in which some of the allied concepts such as *separable*, or *(bi)extensional*, can be conveniently incorporated when technically necessary. Via the pairing of objects with attributes, Chu spaces comprise a means of categorizing objects by their attributes, where categorization implements the satisfaction relation r. Chu spaces can be generalized to multi-valued satisfaction relations, such as relations satisfied in terms of a (conditional) probability (again see [10, 11, 16–19, 22] for further examples and applications).

Let us give a very basic example of a Chu space in terms of matrix arrays as they arise very frequently in a wide range of computation [18, Chap. 1].

Example B.1.1 With $K = 2 = \{0, 1\}$, we take the Objects to consist of a set $X = \{a, b, c\}$. This can be represented as the Chu space

$\Vdash$								
a	0	1	0	1	0	1	0	1
b	0	0	1	1	0	0	1	1
c	0	0	0	0	1	1	1	1

where for Attributes, we could take any 8-membered set A to index the columns. As pointed out in [18, Chap. 1], it is convenient to view the columns as self-identifying, with each column a function $X \to K$, otherwise expressed as $A \subseteq K^X$ (Chu spaces so organized, with $\Vdash$ understood, are called *normal*). Object and Attributes become Tokens and Types, respectively in [21]. This example shows how Chu spaces (and Classifiers) naturally follow from matrix arrays as, for instance, the latter result from encoding of bit strings, This methodology is treated systematically and extensively in [16, 18, 19], for instance.

The arrows in the category **Chu** are prescribed as follows:

Definition B.1.2 A *Chu morphism (transform)* of a Chu space $\mathcal{A} = (A, r, X)$ to a Chu space $\mathcal{B} = (B, s, Y)$, is a pair of functions $(f, \bar{f})$ with $f : A \longrightarrow B$, and $\bar{f} : Y \longrightarrow X$, such that for all $a \in A$, and $y \in Y$, we have $s(f(a), y)$ if and only if $r(a, \bar{f}(y))$.

Chu morphisms are manifestly equivalent to the *infomorphisms* of [21].

B.2   Example: The Dirac-Von Neumann Formulation of QM as a Chu Space

This example follows from [23, Sect. 3]. Take some system, and let S denote its set of *states*, and Q the set of *questions* that can be asked of the system. Usually only "yes/no" questions will be asked, but in general the result of asking a question in a given state is *probabilistic*, as we have seen on a number occasions throughout. Formally, this can be represented by an evaluation $e : S \times Q \longrightarrow [0, 1]$, where $e(s, q)$ is the probability that the question q will receive the answer "yes" when the system is in state S.

Let $\mathcal{H}$ be a complex Hilbert space. The Chu space in question is $(\mathcal{H}_0, \mathbf{L}(\mathcal{H}), e_{\mathcal{H}})$, where:

(1) **Objects**: $\mathcal{H}_0 = \mathcal{H}/0$, the set of non-zero vectors in $\mathcal{H}$.
(2) **Attributes**: $\mathbf{L}(\mathcal{H})$ - the lattice of closed subspaces of $\mathcal{H}$, the standard notion of "yes/no" questions in QM. The observable corresponding to the subspace S, is the self-adjoint operator whose spectral decomposition satisfies $S \oplus S^{\perp} \cong \mathcal{H}$. To each subspace S, there corresponds the projection P_S.
(3) The evaluation $e_{\mathcal{H}}$ is the fundamental formula for providing the basic predictive content of QM:

$$e_{\mathcal{H}}(\psi, S) = \frac{\langle \psi | P_S \psi \rangle}{\langle \psi | \psi \rangle} = \frac{\langle P_S \psi | P_S \psi \rangle}{\langle \psi | \psi \rangle} = \frac{\| P_S \psi \|^2}{\| \psi \|^2}$$

(it can be checked that this is equivalent to working with normalized vectors only).

B.3   Algebraic Operations on Chu Spaces

Pratt in [17, 19] established three important operations on Chu spaces as they are instrumental in the theory of higher dimensional automata: *concurrence, orthocurrence* and *choice*. These we recall as follows:

1. Consider two Chu spaces $\mathcal{A} = (A, r, X)$ and $\mathcal{B} = (B, s, Y)$. The *asynchronous parallel composition*, or *concurrence*, $\mathcal{A} \| \mathcal{B}$ is defined as $(A + B, t, X \times Y)$, where $t(a, (x, y)) = r(a, x)$ and $t(b, (x, y)) = s(b, y)$. The operational sense of this relation is the process whereby $\mathcal{A}$ and $\mathcal{B}$ occur independently of each other.

2. The tensor product, or *orthocurrence*, $\mathcal{A} \otimes \mathcal{B}$ of two processes $\mathcal{A} = (A, r, X)$ and $\mathcal{B} = (B, s, Y)$, is defined as $(A \times B, t, Z)$, where Z denotes the set of Chu transforms (morphisms) from $\mathcal{A}$ to the transpose $\mathcal{B}^{\perp} = (Y, \hat{r}, B)$ of $\mathcal{B}$ ($\hat{r}(y, b) = r(b, y)$), and $t((a, b), (f, g)) = s(b, f(a))$. This provides a sense of "flow through" a given system. For instance, we might consider a network $\mathcal{A}$ of trains passing through a network $\mathcal{B}$ of stations; or more simply, a system of signals flowing through a gated circuit, and likewise, two colliding particles. As for $\mathcal{B} \otimes \mathcal{A}$, this follows from the appropriate replacement in the definition, and up to bijection, $\mathcal{A} \otimes \mathcal{B}$ and $\mathcal{B} \otimes \mathcal{A}$ are isomorphic as Chu spaces; technically, they have the same carrier, cocarrier and matrix [17, 19].

3. The *choice* $\mathcal{A} \sqcup \mathcal{B}$ of two processes $\mathcal{A} = (A, r, X)$ and $\mathcal{B} = (B, s, Y)$ is defined as $(A + B, t, X + Y)$ where $t(a, x) = r(a, x)$, $t(b, y) = s(b, y)$, and $t(a, y) = t(b, x) = 0$. This describes how the events of $\mathcal{A} \sqcup \mathcal{B}$ are formed as the disjoint union of those of $\mathcal{A}$ and $\mathcal{B}$, prescribing a partition into two kinds: (i) those assigning 0 to all events of $\mathcal{B}$ choosing to execute $\mathcal{A}$, any state allowable in $\mathcal{A} \sqcup \mathcal{B}$, and conversely (ii) choosing $\mathcal{B}$, and then setting all events of $\mathcal{A}$ equal to 0.

These operations can be applied directly to the classifiers in the isomorphic category **Chan** featuring in Chap. 3, when the Chu space relations r, s are replaced by the classifications $\Vdash_\mathcal{A}$, $\Vdash_\mathcal{B}$, respectively.

Theories and Local Logics in Channel Theory

C

C.1   Theory in Information Flow

Here we will summarize the essential concepts and results pertaining to theories, local logics, and the logic infomorphisms of Channel Theory, referring to [21] for further details (see also [22, 24]). We commence with some set Ξ, which could be viewed as set of "Types", with a (binary) consequence relation $\vdash$ between subsets of Ξ. A *sequent* is a pair $I = \langle \Gamma, \Delta \rangle$ of subsets of Ξ (it is sometimes useful to view Γ and Δ as sets of *situation types*). A sequent $I = \langle \Gamma, \Delta \rangle$ is said to *hold of a situation s* provided that, if s supports every type in Γ, then it supports some type in Δ.

Definition C.1.1 A *Theory* is a pair $T = \langle \Xi, \vdash_T \rangle$, where $\vdash_T$ is a consequence relation on Ξ. A *constraint* of the theory T is a sequent $\langle \Gamma, \Delta \rangle$ of Ξ for which $\Gamma \vdash_T \Delta$. A a sequent $\langle \Gamma, \Delta \rangle$ is said to be *T-consistent* if $\Gamma \nvdash_T \Delta$.

Example C.1.1 We can exemplify the sequent concept in a straightforward case. Suppose Ξ is the set of polynomials in two variables x, y, and let $\vdash$ be the consequence relation consisting of sequents $\langle \Gamma, \Delta \rangle$, such that each pair $(u, v) \in \mathbb{R}^2$ satisfying all equations in Γ, satisfies some equation in Δ. For instance, a constraint of the theory might be $x^2 + y^2 = 25$, $3x = 4y \vdash x = \pm 4$ [21, Example 9.2, p.118].

In practice, conceiving some aspects of a situation either causally requiring, or merely causally allowing, other aspects of a situation, makes the above definition clear.

In fact, any classifier has its own associated theory seen as follows [21, Proposition 9.5]:

© The Editor(s) (if applicable) and The Author(s), under exclusive license to Springer Nature Switzerland AG 2026
C. Fields and J. Glazebrook, *Distributed Information and Computation in Generic Quantum Systems*, Synthesis Lectures on Engineering, Science, and Technology,
https://doi.org/10.1007/978-3-031-97263-8

Definition C.1.2 A *Theory* $\mathsf{Th}(\mathcal{A}) = (\Sigma_\mathcal{A}, \vdash_\mathcal{A})$ *generated by a classification* $\mathcal{A}$, is one that satisfies for all types α and all sets Γ, Γ', Δ, Δ', Σ', Σ_0, Σ_1 of types:

(1) *Identity*: $\alpha \vdash \alpha$.
(2) *Weakening*: If $\Gamma \vdash \Delta$, then $\Gamma, \Gamma' \vdash \Delta, \Delta'$.
(3) *Global cut*: If $\Gamma, \Sigma_0 \vdash \Delta, \Sigma_1$, for each partition $\langle \Sigma_0, \Sigma_1 \rangle$ of Σ, then $\Gamma \vdash \Delta$.

More generally, any theory $T = \langle \Sigma, \vdash_T \rangle$ is said to be *regular* if it satisfies the above three conditions.

Similar to how classifiers can admit infomorphisms between themselves, Theories have their own notion of morphism; namely, a *Theory interpretation* [21, Sect. 9.4]. Given a Theory T, let $\mathsf{Typ}(\mathsf{T})$ be its set of types, and $\vdash_T$ its consequence relation.

Definition C.1.3 A *(regular) Theory interpretation* $f : T_1 \longrightarrow T_2$, is a function from $\mathsf{Typ}(\mathsf{T}_1)$ to $\mathsf{Typ}(\mathsf{T}_2)$, such that for each $\Gamma, \Delta \subseteq \mathsf{Typ}(\mathsf{T}_1)$, if $\Gamma \vdash_{T_1} \Delta$, then $f[\Gamma] \vdash_{T_2} f[\Delta]$ (here "$f[\ ]$" denotes the image of the theory under the function).

C.2   Local Logic

Definition C.2.1 A *local logic* consists of a triple $(\mathcal{L} = \langle \mathsf{Tok}(\mathcal{L}), \mathsf{Typ}(\mathcal{L}), \Vdash_\mathcal{L} \rangle, \vdash_\mathcal{L}, \mathsf{N}_\mathcal{L})$ in which we have:

(1) a classifier $\mathcal{L} = \langle \mathsf{Tok}(\mathcal{L}), \mathsf{Typ}(\mathcal{L}), \Vdash_\mathcal{L} \rangle$ (note that the classifier is the first term in the triple),
(2) a regular theory $\mathsf{Th}(\mathcal{L}) = (\mathsf{Typ}(\mathcal{L}), \vdash_\mathcal{L})$, and
(3) a subset $\mathsf{N}_\mathcal{L} \subset \mathsf{Tok}(\mathcal{L})$, called *the normal tokens* of $\mathcal{L}$, which are tokens that satisfy all of the constraints of the theory $\mathsf{Th}(\mathcal{L})$ in (2).

For any classifier $\mathcal{A}$ there is the natural local logic $\mathsf{Lg}(\mathcal{A})$ generated by $\mathcal{A}$ itself: it has the classifier $\mathcal{A}$, a regular Theory $\mathsf{Th}(\mathcal{A}) = (\mathsf{Typ}(\mathcal{A}), \vdash_\mathcal{A})$, and all of its tokens are normal. For any local logic $\mathcal{L}$ on $\mathcal{A}$, we have $\mathcal{L} = \mathsf{Lg}(\mathcal{A})$ by [21, Proposition 12.7]. Intuitively, a local logic is "local" to the classifier that generates it. Infomorphisms provide the mapping of the local logic of one classifier to that of another; hence, we can think of channels as supporting the flow of locally-defined logical relations between classifiers. In [22] it was noted that any classifier can be interpreted as defining a coarse-graining, and hence a scale at which information is being organized and represented. Thus, each local logic can be thought of as a logic at some level of description.

C.3   Logic Infomorphisms

In [25], $\mathcal{L}$ is called an *information context*, and $\vdash$ is a binary relation relating sets of situation types. In this case, $N_{\mathcal{L}}$ is said to be a set of *normal situations*. Intuitively, these are the situations about which the available information is concerned. They may comprise all, or only some, of the situations satisfying that information. For instance, we may start with some set of normal situations accounting for an individual's experiences to date, and then the information context consists of all the sequents satisfied by this experience, and consistent with it. Stepping outside of the context generates "surprise" in the sense of an expectation violation (c.f. [26]). These observations motivate asking to what extent an infomorphism between classifiers (denoted Cl) will respect their respective local logics. This is specified by the following: ([21, 12.3]).

Definition C.3.1 A *logic infomorphism* $f : \mathcal{L}_1 \leftrightarrows \mathcal{L}_2$, consists of a covariant pair $f = \langle f^{\smallfrown}, f^{\vee} \rangle$ of functions satisfying:

(1) $f : \mathrm{Cl}(\mathcal{L}_1) \leftrightarrows \mathrm{Cl}(\mathcal{L}_2)$ is an infomorphism of classifiers.
(2) $f^{\smallfrown} : \mathrm{Th}(\mathcal{L}_1) \longrightarrow \mathrm{Th}(\mathcal{L}_2)$ is a (regular) theory interpretation, and
(3) $f^{\vee}[N_{\mathcal{L}_2}] \longrightarrow N_{\mathcal{L}_1}$ (Comment: the notation is that of [21, Definition 12.16]. The "[ ]" notation is explained in Definition C.1.3 above. By covariance of $f^{\vee}$, (3) is equivalent to the pushforward $f^{\vee}(N_{\mathcal{L}_2}) \subseteq N_{\mathcal{L}_1}$.)

Consequences of this definition are outlined in [21, Sect. 12.3], in which we underscore the fact that the flow of information through a network of logic infomorphisms is naturally interpretable as "inference" in the usual sense. As the classification $\Vdash$ can be considered time- and context-dependent, these inferential processes can be regarded as having similar dependencies.

Motivated by Situation Theory, the following are introduced in [24, Sect. 2.1]. Let $[\Gamma; \Delta]$ denote the set of tokens a which are positively described by Γ, and negatively by Δ. This means that $a \Vdash_A \alpha$, for every $\alpha \in \Gamma$, and $a \nVdash_A \alpha$, for every $\alpha \in \Delta$. Elements of $[\Gamma; \Delta]$ are counter examples to the inference from Γ to Δ. Thus, $\Gamma \vdash_A \Delta$, if and only if $[\Gamma; \Delta] = \emptyset$. Accordingly:

Definition C.3.2 A local logic $\mathcal{L}$ is i) *sound* if $\Gamma \vdash_{\mathcal{L}} \Delta$, then $N_{\mathcal{L}} \cap [\Gamma; \Delta] = \emptyset$, and ii) is *complete* if $N_{\mathcal{L}} \cap [\Gamma; \Delta] = \emptyset$, then $\Gamma \vdash_{\mathcal{L}} \Delta$.

Local logics and logic infomorphisms contribute significantly to the logical structure of the distributed systems of information flow, and logical gating thereof, as we have introduced these throughout various chapters.

Sheaves and Toposes

D

D.1 Presheaf and Sheaf—Definitions

The concept of a (pre)sheaf appears in most graduate level textbooks in algebraic geometry (for instance [27, 28]). We will recall the basic concepts.

Let X be a topological space, and suppose that a certain set $\mathcal{F}(U)$ is associated with every open set U in X, and for any pair of open sets $U \subseteq V$ there exists a mapping:

$$r_U^V : \mathcal{F}(V) \longrightarrow \mathcal{F}(U) \tag{D.1}$$

This system is called a *presheaf*, denoted $\mathcal{F}$, provided the following conditions are satisfied:

1. If U is empty, then $\mathcal{F}(U)$ consists of a single element.
2. r_U^U is the identity mapping.
3. For any triple of open sets $U \subseteq V \subseteq W$, we have $r_U^W = r_U^V \circ r_V^W$ ('restriction').

Often $\mathcal{F}(U)$ is referred to as the *section* of $\mathcal{F}$ over U. When, for instance, all the sets $\mathcal{F}(U)$ are groups, or modules over a ring, or rings themselves, we can speak of a presheaf of groups, etc.

The presheaf $\mathcal{F}$ on X becomes a *sheaf* if, for every open set $U \subset X$, and for every open covering of it $U = \bigcup U_\alpha$, the following conditions are satisfied:

© The Editor(s) (if applicable) and The Author(s), under exclusive license to Springer Nature Switzerland AG 2026
C. Fields and J. Glazebrook, *Distributed Information and Computation in Generic Quantum Systems*, Synthesis Lectures on Engineering, Science, and Technology,
https://doi.org/10.1007/978-3-031-97263-8

1. If $r^U_{U_\alpha} s_1 = r^U_{U_\alpha} s_2$, for $s_1, s_2 \in U$ and all U_α, then $s_1 = s_2$. In particular, if $s \in F(U)$ is such that $s|U_\alpha = 0$, for all α, then $s = 0$.
2. If $s_\alpha \in \mathcal{F}(U_\alpha)$ are such that

$$r^{U_\alpha}_{U_\alpha \cap U_\beta} s_\alpha = r^{U_\beta}_{U_\alpha \cap U_\beta} s_\beta$$

then there exists an $s \in \mathcal{F}(U)$ for which $s_\alpha = r^U_{U_\alpha} s$, for all U_α.

D.2   Sites and the Topos Concept

A. Grothendieck, in his Bois-Marie seminars of the 1960s, proposed a generalization of a sheaf on a topological space by introducing the concept of a *sheaf on a site*. This was a further step in abstracting algebraic-geometric concepts towards achieving a general framework for his then current research, eventually leading to the acclaimed concept of a *topos* [29]. We will provide a broad explanation of terms following [30]. To commence, consider a category $\mathfrak{C}$ having finite projective limits (viz inverse limits $\varprojlim$; see [4, Chap. 5] and [2]). For each object U of $\mathfrak{C}$, a *Grothendieck topology* on $\mathfrak{C}$ is a way to specify families of maps $(U_i \longrightarrow U)$, suitably indexed, termed *covering families*. The latter are analogous to an open covering of an open subset of a topological space that satisfy certain properties (see [29, 30]). Once selecting a topology on $\mathfrak{C}$, the latter becomes a *site*, and a sheaf of sets on $\mathfrak{C}$ is definable in the same way as when $\mathfrak{C}$ is the category of open sets of a topological space. Keeping in mind how the concept of sheaf evolved in Sect. D.1 above, specifically, a *sheaf of sets* $\mathcal{F}$ on $\mathfrak{C}$ is a contravariant functor $U \longrightarrow \mathcal{F}(U)$ on $\mathfrak{C}$, with the property that for any covering family $U_i \longrightarrow U)$, a section $s \in \mathcal{F}(U)$ can be identified via the restriction maps with a family of sections s_i of $\mathcal{F}$ on the U_i's that coincides with $U_i \times_U U_j$ (the 'intersections'). This leads to defining a *topos* T as a category of sheaves of sets on a site $\mathfrak{C}$, which is then a *defining set* for T.

The intrinsic properties of a topos T are discussed in [30] in light of how Grothedieck intended them to be used for a general framework that combined number theory with algebraic geometry [29]. The topos concept has since been applied extensively to a range of mathematical ideas, particularly those allied to logic and structure [31, 32], as well as applications to topics we have discussed previously in various places; these topics include nonlocality/contextuality [33], and institutional information flow in logical environments [34].

Rice's theorem [35] states that only trivial properties of computing programs are algorithmically decidable, i.e. decidable in finite time by a Turing machine with arbitrarily-large memory.

We begin by distinguishing the class **RE** of recursively-enumerable languages from the smaller class **R** of recursive languages. A language $\mathscr{L}$ is in **RE** if a Turing machine can decide in finite time if a given finite string s is in $\mathscr{L}$. A language $\mathscr{L}$ is in **R** if a Turing machine can decide in finite time if a given finite string s is or is not in $\mathscr{L}$. See [36] for background and details.

The Halting Problem is the problem of deciding, for any program (language) $\mathscr{L}$ and any input string s whether s is in $\mathscr{L}$, i.e. whether the program $\mathscr{L}$ "halts" by outputting either '1' (or "accept" indicating $s \in \mathscr{L}$) or '0' (or "reject" indicating $s \notin \mathscr{L}$). This problem is undecidable [37]; hence **RE** $\supset$ **R**. Again see [36] for background and details.

We can state Rice's theorem informally as:

Theorem E.0.1 (Rice [35] (informal)) *Whether an arbitrarily-chosen program P implements a nontrivial function $y = f(x)$ is (finite Turing) undecidable.*

And prove it by showing that it reduces to the Halting problem.

Proof (Sketch) If one can decide that an arbitrary P computes $y = f(x)$, one can decide whether the program P' specified by:

© The Editor(s) (if applicable) and The Author(s), under exclusive license to Springer Nature Switzerland AG 2026
C. Fields and J. Glazebrook, *Distributed Information and Computation in Generic Quantum Systems*, Synthesis Lectures on Engineering, Science, and Technology,
https://doi.org/10.1007/978-3-031-97263-8

$$P \leftarrow x$$

$$\text{If } (y = f(x)), \text{ Halt}$$

halts. But whether an arbitrary P' halts is undecidable [37].   □

More formally, we can define a "property" P as a set of languages, i.e. say that any language $\mathscr{L}$ that satisfies the property P, belongs to P, namely $\mathscr{L} \in P$. A property P is *trivial* if: (i) P is not satisfied by any $\mathscr{L} \in \mathbf{RE}$; or (ii) P is satisfied by all $\mathscr{L} \in \mathbf{RE}$. Conversely, a property is *non-trivial* if it is satisfied by some $\mathscr{L}_i \in \mathbf{RE}$, and is not satisfied by some $\mathscr{L}_j \in \mathbf{RE}$, $i \neq j$. Therefore, for a non-trivial property P, there exist Turing machines M_i and M_j such that M_i has P but M_j does not.

With these definitions, we can state the theorem formally:

Theorem E.0.2 (Rice [35]) *If P is a non-trivial property of an* **RE** *language, then:*

$$\mathscr{L}_P = \{\langle M \rangle | \mathscr{L}(M) \in P\} \tag{E.1}$$

is undecidable.

The following proof is adapted from [36].

Proof Consider a property P such that the empty language $\emptyset \notin P$; the proof for the case of $\emptyset \in P$ is analogous. Since P is nontrivial, $\exists \mathscr{L} \neq \emptyset, \mathscr{L} \in P$. Let $M_{\mathscr{L}}$ be a Turing machine that accepts $\mathscr{L}$, i.e. accepts all strings in $\mathscr{L}$.

Let M be an arbitrary Turing machine, let w be an arbitrary string, and let $x \in \mathscr{L}$. We now construct a Turing machine M' that performs the following operations:

- M' simulates M acting on the input w;
- if the simulation of M acting on w halts on 'reject', M' halts;
- if the simulation of M acting on w halts on 'accept', M' reads x as input;
- if M' reads x as input, M' simulates $M_{\mathscr{L}}$ on x.

We can now observe:

- if M rejects w, M' never reads x; hence the language accepted by M' is $\emptyset$. Since $\emptyset \notin P$, M' does not have P.
- if M accepts w, M' accepts x. By definition, x is an arbitrary string in $\mathscr{L}$; hence M' accepts $\mathscr{L}$. By definition, $\mathscr{L} \in P$, so M' does have P.

But M and w are arbitrary, so M' both has and does not have P. Contradiction.   □

References

1. Eilenberg, S. and Mac Lane, S. (1945). Relations between homology and homotopy groups. *Ann. Math.* 46, 1945, 480–509.
2. Mac Lane, S. *Categories for the Working Mathematician.* Graduate Texts in Mathematics. Springer, Heidelberg-New York, 1972.
3. Adámek, J., Herrlich, H. and Strecker, G. E. *Abstract and Concrete Categories: The Joy of Cats.* Wiley, New York, 2004. Available at http://katmat.math.uni-bremen.de/acc (Accessed May 26, 2019).
4. Awodey, S. *Category Theory.* (Oxford Logic Guides, 62). Oxford, UK: Oxford University Press, 2010.
5. Brown, R. *Topology and Groupoids.* Book Surge Publishing, Charleston SC, 2006.
6. Baianu, I. C., Brown, R., Georgescu, G. and Glazebrook, J. F. Complex nonlinear biodynamics in categories, higher dimensional algebra and Lukasiewicz-Mosil Topos: Transformations of neuronal, genetic and neoplastic networks. *Axiomathes* 16 (2206), 65–122.
7. Brown, R. and Porter, T. (2003). Category theory and higher dimensional algebra: Potential descriptive tools in neuroscience. In: Singh, N. (Ed.) Proceedings of the International Conference on Theoretical Neurobiology, Delhi, February 2003, National Brain Research Centre (pp. 80–92), 2003.
8. Goguen, J. A. (1991). A categorical manifesto. *Mathematical Structures in Computer Science* 1, 49–67.
9. Ehresmann, A. C. and Vanbremeersch, J.-P. *Memory Evolutive Systems; Hierarchy, Emergence, Cognition (Studies in Multidisciplinarity).* Elsevier, New York, 2007.
10. Fields, C. and Glazebrook, J. F. A mosaic of Chu spaces and Channel Theory II: Applications to object identification and mereological complexity. *Journal of Experimental and Theoretical Artificial Intelligence* 31(2) 2019, 237–265.
11. Fields, C.; Glazebrook, J. F. Information flow in context-dependent hierarchical Bayesian inference. *J. Expt. Theor. Artif. Intell.* 34 2022, 111–142.
12. Fields, C.; Friston, K.J.; Glazebrook J.F.; Levin M. A free energy principle for generic quantum systems. Prog. Biophys. Mol. Biol. 2022, 173, 36–59.

© The Editor(s) (if applicable) and The Author(s), under exclusive license to Springer Nature Switzerland AG 2026

C. Fields and J. Glazebrook, *Distributed Information and Computation in Generic Quantum Systems*, Synthesis Lectures on Engineering, Science, and Technology, https://doi.org/10.1007/978-3-031-97263-8

13. Fields, C.; Glazebrook, J. F. Separability, contextuality, and the quantum Frame Problem. International Journal of Theoretical Physics 2023; 62: 159.

14. Barr, M. (1979). *-Autonomous categories, with an appendix by Po Hsiang Chu. *Lecture Notes in Mathematics* 752. Berlin: Springer.

15. Barr, M. (1991). *-Autonomous categories and linear logic. *Mathematical Structures in Computer Science* 1 1991, 159–178.

16. Pratt, V. (1995). Chu spaces and their interpretation as concurrent objects. *Lecture Notes in Computer Science* 1000, 392–405.

17. Pratt, V. (1997). Types as Processes, via Chu spaces. Invited paper. Proceedings 'Express'97: Fourth Workshop on Expressiveness in Concurrency'. Santa Margherita, Italy. September 1997. *Electronic Notes in Theoretical Computer Science* 7 21pp. http://www.elsevier.nl/locate/entcs/volume7.html

18. Pratt, V. (1999). Chu spaces. *School on Category Theory and Applications (Coimbra 1999)*, Vol. 21 of *Textos Mat. Sér. B*, University of Coimbra, Coimbra. (pp. 39–100).

19. Pratt, V. Higher dimensional automata revisited. *Mathematical Structures in Computer Science* **10**(4) (2000), 525–548.

20. Gupta, V. Chu spaces: A Model for Concurrency. PhD Thesis, Stanford Univ. CA, 1994.

21. Barwise, J. and Seligman, J. (1997). *Information Flow: The Logic of Distributed Systems* (Cambridge Tracts in Theoretical Computer Science, 44). Cambridge, UK: Cambridge University Press.

22. Fields, C. and Glazebrook, J. F. (2019a). A mosaic of Chu spaces and Channel Theory I: Category-theoretic concepts and tools. *Journal of Experimental and Theoretical Artificial Intelligence* 31(2), 177–213.

23. Abramsky, S. Big toy models: Representing physical systems as Chu spaces. *Synthese* 186 n(2012), 697–718.

24. Seligman, J. Situation Theory Reconsidered. In (Baltag, A. and Smets, S.) Johan van Benthem on Logic and Information Dynamics. Outstanding Contributions to Logic, vol. 5, pp. 895–932. Springer, Cham, 2014.

25. Barwise, J. (1997). Information and Impossibilities. *Notre Dame Journal of Formal Logic* 38(4), 488–515.

26. Friston, K. The free-energy principle: A unified brain theory? Nat. Rev. Neurosci. 2010, 11, 127–138.

27. Hartshorne, R. *Algebraic Geometry*. Graduate Texts in Mathematics 52. Springer-Verlag, New York NY, 1977.

28. Shafarevich, I. R. *Basic Alegbraic Geometry*. Grundlehren 213. Springer-Verlag, Heidelberg, 1974.

29. Artin, M., Grothendieck, A. and J. L. Verdier. [SGA 4] Théories des topos et cohomologie étale des schémas. *Séminairede Géometrie Algébraic du Bois-Marie 1963-64* SLN 269, 270, 305, Springer-Verlag Berlin, Hedielberg 1972, 1973.

30. Illusie, L. What is ... a Topos? *Notices of the Amer. Math. Soc.* 51(9) (2004), 1060–1061.

31. Mac Lane, S. and Moerdijk, I. *Sheaves in Geometry and Logic - A First Introduction to Topos Theory*. Springer Science + MediaBusiness, New York, 1992.

32. Landsman, K. Topos theory and quantum logic. In *Foundations of Quantum Theory. Fundamental Theories of Physics* 188, pp. 459–494. Springer, Cham, 2017.

33. Butterfield, J. and Isham C. J. Topos perspective on the Kochen-Specker theorem: I. Quantum states as generalized valuation. *Int. J. Theor. Phys.* 37 1998, 2669–2733.

34. Kent, R. E. Conceptua: Institutions in a topos. arXiv1811.02041v1 [cs.LO] (2018).

35. Rice, H. G. Classes of recursively enumerable sets and their decision problems. *Trans. Amer. Math. Soc.* 74 (1953), 358–366.

36. Hopcroft, J. and Ullman, J. *Introduction to Automata Theory, Languages, and Computation*. Addison-Wesley, Boston MA, (1979).
37. Turing, A. On computable numbers, with an application to the Entscheidungsproblem. *Proc. London Math. Soc. Ser. 2* 42 (1937), 230–265.
38. Feynman, R. P. (1987). Negative probability. In: Hiley, B. J. and Peat, F. D. (Eds.) *Quantum Implications: Essays in Honor of David Bohm*. London: Routledge and Kegan Paul (pp. 235–248).
39. Fields, C. and Glazebrook, J. F. Do Process-1 simulations generate the epistemic feelings that drive Process-2 decision making? *Cognitive Processing* 21 2020, 533–553.
40. Pezzulo, G., Parr, T. and Friston, K. The evolution of brain architectures for predictive coding and active inference. *Phil. Trans. R. Soc. B* 377, 20200531.
41. Shanahan, M. and Baars, B. Applying global workspace theory to the frame problem. *Cognition* 98 2005, 157–176.

Zeitfracht Medien GmbH
Ferdinand-Jühlke-Straße 7
99095 Erfurt, Deutschland
produktsicherheit@kolibri360.de